the EARTH in TURMOIL

EARTHQUAKES, VOLCANOES, AND THEIR IMPACT ON HUMANKIND

KERRY SIEH
SIMON LEVAY

W. H. Freeman and Company
New York

Cover designer: Pixel Press
Text designer: Cambraia Magalhaes
Cover image: Freeway connector that collapsed in the 1994 Northridge earthquake.
(Jim Dewey, USGS)

Library of Congress Cataloging-in-Publication Data
Sieh, Kerry E.
    The earth in turmoil : earthquakes, volcanoes, and their impact on humankind /
Kerry Sieh and Simon LeVay.
        p.   cm.
    Includes bibliographical references and index.
    ISBN 0-7167-3151-7
    ISBN 0-7167-3651-9 (paperback)
    1. Earthquakes—United States.   2. Volcanoes—United States.
I. LeVay, Simon.   II. Title.
QE535.2.U6S585   1998
551.22′0973—dc21   98-16882
                                                                                              CIP

Printed in the United States of America

First paperback printing 1999

CONTENTS

PREFACE

The Earth in Turmoil is the product of a collaboration between two scientists of quite different backgrounds. Kerry Sieh, a geologist at the California Institute of Technology (Caltech), is best known for his research on the San Andreas fault, but he has also studied many other earthquake faults around the world and has done research on volcanoes too. Simon LeVay is a brain scientist by training; at the start of this project, he knew next to nothing about earthquakes or volcanoes. Thus, between us, we may claim to have two qualities important to writing a popular science book: an insider's and an outsider's view of the subject matter.

Kerry Sieh's research is discussed in some chapters—a circumstance that would ordinarily call for first-person narrative. Because the book has two authors, however, and the English language lacks a set of pronouns to refer to one of two speakers, we refer to him in the third person. This convention should not be taken as an indication that the book was written by Simon LeVay about Kerry Sieh. It was a collaboration— the brainchild of two parents.

The book has godparents, too: the many people who helped it along by talking with us about their own work or experiences or by reading portions of the book. These include geologists and geophysicists Brian Atwater (University of Washington); Robert Duncan (Oregon State University, Corvallis); Charlie Bacon, Bill Ellsworth, Ruth Harris, and Dave Hill (USGS, Menlo Park); Keith Kelson (William Lettis and Associates, Walnut Creek); Steve Ward (University of California, Santa

Cruz); Dave Clague (Monterey Bay Aquarium Research Institute); Tom Heaton and Brian Wernicke (Caltech); Roy Van Arsdale (University of Memphis); Tony Crone (USGS, Denver); John Ebel (Boston College); Pradeep Talwani (University of South Carolina, Columbia); Nano Seeber (Lamont-Doherty Earth Observatory, Palisades, New York); Christina Heliker and Paul Okubo (Hawaii Volcano Observatory); and Alex Malahoff (University of Hawaii, Honolulu). We are also indebted to structural engineers and seismic-safety administrators Earl Schwartz, Ben Schmidt, Karl Deppe, Scott McGill, and Larry Brugger; Mammoth Lakes residents Glenn Thompson, Nancy O'Kelly, and Sam Walker; Michael Rutter of MIT Press, and our friend Laurie Saunders. We thank all of them.

We are especially grateful to four people who read the entire manuscript and provided us with many helpful suggestions. They are Rick Hazlett (Pomona College), Kate Hutton (Caltech), Bettyann Kevles, and our editor at W. H. Freeman and Company, Jonathan Cobb.

the EARTH in TURMOIL

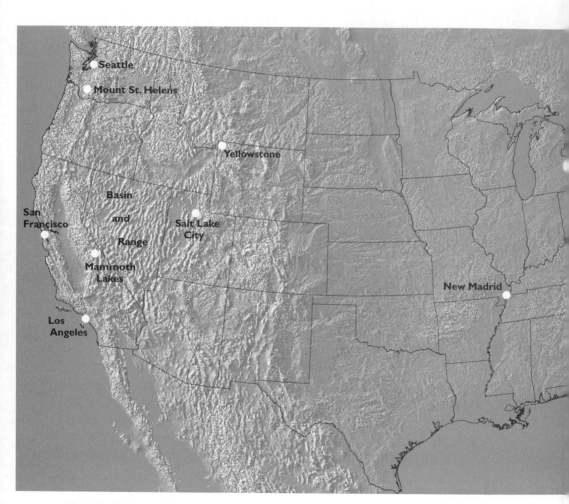

Shaded-relief map of the contiguous United States showing the locations visited in this book.

INTRODUCTION

Early on the morning of October 28, 1983, Lawana and William Knox set out from their home in Challis, Idaho, for a day's elk-hunting. By six minutes after eight, Lawana was sitting in Arentson Gulch, a dry streambed in the foothills of the Lost River Range. With her gun by her side, she watched a few deer grazing quietly on a nearby hillside. Her husband, meanwhile, was driving a herd of elk down the slope toward her.

At that moment, both Lawana and William became aware of an unusual rumbling. Soon they felt the ground trembling, and within a few seconds the motion was so violent that William, fearing a heart attack, threw himself to the ground. Lawana, already on the ground, felt as if her head were being shaken off her shoulders. The motion lasted only about thirty seconds, "but it felt like a lifetime," Lawana said. "No building could have taken the shaking," added William.

Many people have felt the violence of a strong earthquake, but what happened next has been experienced by no more than a handful of people in all of history. Just as the tremors were subsiding, Lawana saw a great rupture tear across the land-

I

scape before her. It was "just as though one took a paintbrush and painted a line along the hill." Within a few seconds, the break had traveled from one horizon to the other. On Lawana's side of the break, the ground dropped; on the far side, it was raised up. The previously smooth-sloped hillside was now interrupted by a five-foot rampart. It extended for miles across Arentson Gulch and, with many zigs and zags, along the base of the Lost River Range, too. The entire range had been lifted up: Borah Peak, which had begun the day at an altitude of 12,662 feet above mean sea level, finished it at 12,662 feet and 5 inches (see Color Plates 1 and 2).

Lawana Knox learned through direct experience something that Earth scientists have had to figure out from years of painstaking fieldwork and laborious analysis: the natural forces that unleash life-threatening events such as earthquakes and volcanic eruptions are the same as those that build much of the landscape around us—not just the Lost River Range but every mountain range, the sunken valleys between them, and the great plateaus and deserts—whole continents, in fact, and the vast hidden floors of the ocean. Knox was witness to one moment in the creation of the world—a creation that did not take place all at once long ago but goes on even today as an inexorable cycle of building up and tearing down.

Earthquakes and volcanoes have fascinated humankind for thousands of years. A part of that fascination, no doubt, is fear—fear of their power to destroy us, or if not to destroy us, then to disrupt our lives greatly. The Borah Peak earthquake did only limited harm: two deaths and modest property damage. Unlike most earthquakes, in fact, it may even have saved a few lives—those of the elk that had been heading unwittingly toward Lawana Knox's rifle. But consider these examples:

• On July 27, 1976, an earthquake only a few times larger than the Borah Peak earthquake leveled Tangshan, a Chinese city east of Beijing. At least a quarter of a million people perished.

• On January 17, 1994, a moderately sized earthquake centered near Northridge, California, did $20 billion worth of damage to buildings, personal property, and infrastructure—and even more if one counts the losses due to the interruption of business or the value of the 33 lives that were lost.

• On May 18, 1980, the eruption of Mount St. Helens, Washington, strewed about a quarter of a cubic mile of the mountain over the landscape. The blast devastated 230 square miles of the surrounding land. Sixty-one people died.

• On August 27, 1883, the volcanic island of Krakatau in Indonesia heralded its obliteration with a blast that was heard 3000 miles away. About 36,000 people were killed by the ensuing eruption and by the accompanying tsunamis (so-called tidal waves). The ejected ash darkened the skies of distant lands.

The cumulative personal and economic devastation caused by earthquakes and volcanoes has never been tallied. But according to official figures, more than a million and a half people have died in earthquakes alone during the twentieth century. The true figures are probably considerably higher. Earthquakes and volcanic eruptions must rank alongside war and plague in the pantheon of human misery.

Fear of earthquakes and volcanoes, and the concomitant desire to mitigate their disastrous consequences, is certainly reason enough to focus our attention on them. But there is another element to our fascination with these cataclysmic events. In the course of our attempt to understand why and how earthquakes and volcanic eruptions occur, we are led into a realm of time, space, and power that dwarfs all human concerns. We embark on a voyage as exciting in its own way as those of Columbus or Magellan, a voyage far across the oceans of time and deep into the Earth's slow-beating heart. From this perspective, humanity is no more than a castaway that clings, for a mere instant, to what it foolishly calls solid ground. Yet, through some curious alchemy, this view of humanity has the power to ennoble rather than to demean: it teaches us to reach beyond our often futile quest for mastery over Nature and to appreciate the terrible beauty of a world that holds the power to destroy us.

The journey of exploration, as laid out in this book, is really twofold. One journey is geographic—a trip across the United States. We start in the Pacific Northwest, travel down to California, and then across the American West to the Mississippi Valley and the East Coast, with a final leap all the way back to Hawaii. This journey illustrates the broad impact earthquakes and volcanoes have had on America's landscape and its inhabitants, from prehistoric times to the present.

The other journey is thematic. Each region that we visit illustrates a particular aspect of our understanding of these geological catastrophes. Along the coast of the Pacific Northwest, we see how the slow but inexorable convergence of two great segments of the Earth—tectonic plates—unleashes the most powerful earthquakes to which our planet is subject. Farther inland, in the Cascade Range, we see how a plate diving into the mantle beneath the North American continent causes volcanic eruptions in the overlying crust. In California, we learn how a sideways sliding motion between two plates—along the notorious San Andreas fault—periodically triggers great earthquakes. As we move eastward, we see how other mechanisms, sometimes poorly understood, cause both earthquakes and volcanoes far from the margins of tectonic plates. Finally, our visit to the Hawaiian Islands illustrates how a mantle plume—a current of hot rock rising from near the Earth's core—raises a procession of towering volcanoes from the deep ocean floor.

Although the examples on which we focus are American, most of the world's great earthquakes and volcanic eruptions—and quite likely those on other planets, too—are caused by one of the mechanisms described in the book. In that sense, the book represents a global, or even a universal, journey.

Let us set out.

1

WHEN PUSH COMES TO SHOVE: GIANT EARTHQUAKES IN THE PACIFIC NORTHWEST

Of the whole circle of earthquake-prone shores that rim the Pacific basin, the 750-mile-long stretch that runs south from British Columbia, past Washington and Oregon to California's Cape Mendocino, has been an exception, a nonperformer, a dud. Elsewhere around the Pacific Rim—in coastal Mexico, Columbia, Chile, New Guinea, Japan, and Alaska—earthquakes of near-incomprehensible magnitude are within human memory. But in the Pacific Northwest, no giant earthquake has occurred in recorded history. In the seismographic stations of Vancouver, Seattle, and Portland, the needles trace out endless smooth spirals on their drums or at most record the tremors of some minor-league, land-based earthquake or the faint echoes of a major shock in some distant, more perilous corner of the Earth. Seismologically, the Cascadia subduction zone, as it is called, is quiet. A little too quiet.

That, at least, is the opinion of Brian Atwater, a U.S. Geological Survey (USGS) scientist at the University of Washington in Seattle. Since 1985, Atwater has been sloshing in rubber boots through Washington's coastal marshes, digging his spade into ancient peat bogs, pulling pieces of drowned trees from their briny graves, and scanning riverbanks and lakeshores for anything that might provide a clue to the mystery of this seismic silence. And clues he has found aplenty: so many, indeed, that the theory he first presented to skeptical colleagues, back in the 1980s, is now well-nigh accepted doctrine. The silence, Atwater has shown, is not the silence of the Earth at rest. It is the silence of a grim struggle between two of Earth's Titans—the North American continent and the floor of the Pacific Ocean—locked in a centuries-old shoving match. Each year the match goes on, and neither side yields. Each year the two combatants strain the harder, until at long last both must blink. And with that blink will come a terrible calamity.

The Deadly Engine

To understand the clues that Atwater has dug up, we must first go back to the beginning of the twentieth century, when another outsider, the German meteorologist and explorer Alfred Wegener (1880–1930), put forward the theory of continental drift. Wegener pointed out (as others, in fact, had done before him) that some of Earth's landmasses look like the disassembled pieces of a jigsaw puzzle. Most strikingly, the long, winding S-shape formed by the Atlantic seaboards of Greenland, North America, and South America could fit fairly snugly against the corresponding western seaboards of Europe and Africa. The fit is even better if one matches not the shorelines but the margins of the continental shelves—the strips of relatively shallow water—that extend variable distances from the coasts into the Atlantic Ocean.

Wegener proposed that all of Earth's continents once formed a single landmass, which he named Pangaea or "all-land." As a result of forces

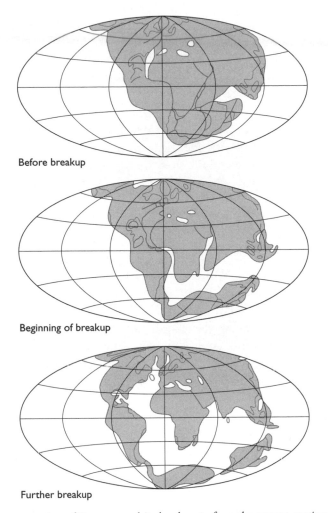

Before breakup

Beginning of breakup

Further breakup

Wegener's reconstruction of Pangaea and its breakup to form the present continents.

that Wegener could not identify, Pangaea broke up. The rupture, he estimated, began about 300 million years ago. Ever since then, the resulting fragments of land have been drifting across Earth's surface as if, like
Wagner's *Flying Dutchman,* they have been condemned to a never-ending quest for anchorage.

Wegener's arguments in support of this proposal went far beyond
the mere coincidence in shape of two continental margins. He also
based his theory on observations in geology, biology, and climatology.
The geological arguments were based on the apparent continuity of

some mountain ranges, coal beds, and other geological structures from one continent to another. If one closes up the Atlantic Ocean, for example, the Appalachian Mountains of the east coast of North America become continuous with Scotland's Grampian Mountains. Similarly, geological formations on the coast of Brazil line up, layer by layer, with equivalent formations on the west coast of Africa.

The biological arguments depended on the existence (either now or in the past) of similar or identical species on more than one continent. Why was it, for example, that the fossilized remains of stegosaurs had been found in places as remote from each other as Texas and India? Didn't this mean that 250 million years ago, when stegosaurs flourished, the world was laid out in such a way that they could walk from one location to the other?

As for climate, Wegener looked at the record preserved in the rocks and the fossils hidden within them. The record was bizarre. In the Mediterranean region, for example, rocks of the Carboniferous period (280 million years ago) contain fossils of tropical species, while in India, rocks of the same age are glacier-gouged. How could India have been so much colder than the Mediterranean if the two regions occupied the same latitudes they do now? Wegener resolved this problem by proposing that India then lay thousands of miles to the south of its present location.

Wegener's theory, attractive through it was, met with great resistance. Neither Wegener nor anyone else could explain how great rocky continents might sail from one part of Earth's surface to another. After all, the continents do not float in the oceans. They are grounded on Earth's mantle, a sphere of solid rock that underlies both land and sea and reaches down nearly 2000 miles to Earth's metallic core. Therefore, most geologists argued, the continents must stay put wherever they were first created. And so, through his lifetime and for 30 years after his death, Wegener's ideas languished on the fringes of respectable geological discourse.

In the 1960s, however, the notion of continental drift suddenly became respectable, even trendy. For now Wegener's ideas were fitted into an even bolder theory, one that presumed to explain the history of Earth's crust over billions of years past and to predict its course for millions of years to come. This theory was *plate tectonics.*

At the root of the theory was a fundamental rethinking of the nature of Earth's mantle. Certainly, the mantle was solid rock. That much was known from its ability to transmit certain types of seismic waves that can travel only in solid matter. But by analyzing seismic waves coming from distant points on the surface of the globe, which travel far beneath Earth's surface to reach the observing station, seismologists found that the mantle is not homogeneous. Rather, there is a layer, between about 100 and 400 miles below the surface, where the waves travel more slowly and die out more rapidly than they do in the layers above or below. This finding suggested that, because of the high pressure and temperature in this zone, the rock is less brittle and more taffylike (that is, more ductile) than the rock overlying it. In addition, seismologists noticed that earthquakes, with certain exceptions, did not occur below a depth of about 100 miles. This finding also could be explained if the lower regions are too soft to undergo the brittle fractures that trigger earthquakes. Thus, geologists came to agree that the crust and the rigid outer part of the mantle—collectively termed the *lithosphere* or "rock sphere"—overlies a more ductile material that, if pushed hard enough, can deform and flow. This layer—named the *asthenosphere* (or "weak sphere")—might yield sufficiently to allow the overlying continents to slide sideways over it.

But if the asthenosphere allowed the continents to be pushed around on Earth's surface, who or what was doing the pushing? The answer to this question came from a whole new realm of exploration—the ocean floor.

Long hidden from human view, the deep ocean floor started to yield its secrets in the 1940s, when mariners discarded their ancient lead lines and began to measure the sea's depth by sonar. The method is simple enough if you possess the right technology. The ship carries a loudspeaker that sends sound waves down into the water. The waves echo off the bottom and travel back to the ship, where they are picked up by a microphone. From the interval between the outgoing sound and the returning echo, and the known velocity of sound in water, the depth can be readily calculated.

By the end of the 1950s, oceanographers using this technique had mapped parts of the oceans and had discovered a whole landscape whose existence had barely been guessed at. In the middle of the Atlantic Ocean, for example, a north-south ridge of undersea mountains rises 6000 to 10,000 feet from the ocean floor. The ridge neatly follows the winding of the Atlantic's eastern and western shores, always staying halfway between the two. It is now known that the Mid-

Atlantic Ridge is just a part of a network of deep-sea ridges that together encircle the globe like the seam on a baseball. Their total length is about 40,000 miles.

Besides the deep-sea ridges, another unsuspected feature of the ocean floor was revealed by the surveys of the 1940s and 1950s. This was a system of trenches—long valleys where the seafloor descended thousands of feet below the surrounding terrain. The Atlantic Ocean has almost no such trenches, but the Pacific Ocean is ringed by a long system of trenches that, for the most part, run parallel to the coasts of the surrounding continents, 70 or more miles offshore.

In 1960 a geology professor at Princeton University, Harry Hess, offered a dramatic synthesis of the oceanographic observations. Hess himself had been a pioneer in the mapping studies; as a naval officer during the Second World War, he had occupied his free time by scanning the deep seafloor. He now suggested that the ridges and trenches were, in effect, the on- and off-ramps of vast conveyer belts that carried large pieces of crust across Earth's surface. New crust, he proposed, was being created continuously by volcanic action at the ridges. This rock moved away from the ridges until it reached a trench, where it sank back down into the mantle—a process called *subduction*. Another geologist, Robert Dietz, added an important detail: it was not just the crust, but the entire lithosphere (crust plus upper mantle) that took part in this cyclical motion.

The 1960s were a decade of great excitement in geology as evidence in support of Hess's theory accumulated. A ship, the *Glomar Explorer,* was specially built to drill into the deep seafloor, and it recovered rock samples consistent with volcanic activity at the mid-ocean ridges. (Its sister ship, the *Glomar Challenger,* once made a pretense of geological exploration while it secretly recovered a portion of a sunken Soviet submarine.) Eventually, submersibles descended to the ridges, giving geologists a direct view of active volcanoes that were erupting basalt—dark, dense, fine-grained lava—onto the seafloor.

But the most exciting evidence in favor of the theory came from the study of the magnetic properties of the seafloor rocks. It was already known, from observations made on land, that when molten rock solidifies it acquires a weak magnetism imposed by Earth's magnetic field. This magnetization occurs because crystals of iron-rich minerals, as they begin to solidify in the fluid magma, can swing freely like compass needles; but as the whole magma hardens, the crystals become locked

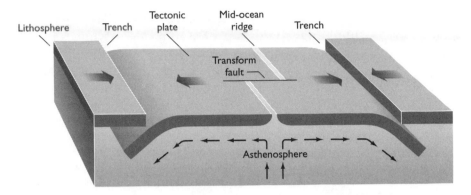

The recycling of the lithosphere. (Adapted from an illustration by B. Isaacs and others)

into position, permanently recording the orientation of Earth's magnetic field as it was at that time.

Studies of the magnetism of ancient volcanic rocks led to a surprising finding: Earth's magnetic poles were not always located where they are now. Even today the position of magnetic north wanders slightly from year to year, so that navigators who use magnetic compasses (if any still do!) have to make annual corrections to get a precise indication of the direction of true north. But over geological time, the north and south magnetic poles have actually exchanged places on many occasions.

During the 1950s, a number of scientists, most notably a group led by Victor Vacquier at Scripps Institute of Oceanography in San Diego, studied the magnetism of the ocean floor. They did this not by drilling rock samples but simply by towing a sensitive magnetometer (a device for measuring the strength and direction of the magnetic field) behind a ship that was traveling across the ocean surface. The magnetometer picked up tiny variations in the magnetism of the volcanic rocks on the seafloor. What the oceanographers found was that, on both sides of a mid-ocean ridge, the rocks were arranged in alternating strips, parallel to one another and to the ridge. In one set of strips, the rocks were magnetized as if Earth's north magnetic pole were in the Arctic (the "normal" arrangement); in the alternating strips, the rocks were magnetized as if the north magnetic pole were in the Antarctic (the "reversed" arrangement). The widths of these strips were quite variable. What was most striking, however, was that the pattern of strips on one side of the ridge exactly mirrored the pattern on the other side (see Color Plate 3).

Three scientists—the Canadian Lawrence Morley, and Frederick Vine and Drummond Matthews of Cambridge University in England—hit upon the correct interpretation of the magnetic strips. As Hess and Dietz had proposed, new lithosphere is continuously being created at the mid-ocean ridges, and it spreads symmetrically away from the ridge in both directions. As new ocean floor is created by the solidification of the volcanic magma, it records the current orientation of Earth's magnetic field. Because Earth's magnetic field "flips" into the reverse orientation every now and then, strips of the floor are alternately imprinted with the normal magnetic pattern and with the reversed pattern.

The magnetic strips thus record the history of the ocean floor, rather in the same way that tree rings record the life history of a tree. But unlike tree rings, which form at the rate of exactly one per year, the intervals between magnetic reversals are variable. So some method was needed to date them. Luckily, such a method had recently been invented. The method measures how much of a radioactive isotope of potassium in the rock, potassium-40, has decayed to a stable isotope of argon, argon-40. Because the half-life of potassium-40 is 1.3 billion years (that is, one-half of a sample of potassium-40 will have decayed in that period), the method can be used to date rocks from any period in Earth's history.

The most recent reversal of Earth's magnetic field took place about 780,000 years ago. The strips of normally magnetized rock that are currently emerging from the Mid-Atlantic Ridge extend about 4.5 miles to the east and the west of the ridge. Thus the average rate of movement has been 4.5 miles divided by 780,000, or about 9 millimeters (about one-third of an inch) per year. (Because the motion is taking place on the surface of a sphere, the rate of motion actually varies along the length of the ridge, being zero at its two ends and highest at its midpoint.) Since the movement is progressing in both directions, the Atlantic Ocean as a whole must be widening at double that rate; that is, by about 1.8 centimeters per year. Working backward in time, and assuming that the rate of movement has always been the same, we can calculate how long ago it was that the Americas first split off from Europe and Africa to form the Atlantic Ocean. It turns out that the South Atlantic opened up about 180 million years ago and the North Atlantic about 100 million years ago. Wegener's original estimate of 300 million years was not too far off the mark. And Columbus had 30 feet less to travel when he crossed the Atlantic 500 years ago than we do today.

One further conceptual step was required to complete the picture. In 1964, the Canadian geophysicist J. Tuzo Wilson suggested that Earth's surface could be divided into a small number of large, rigid plates. Each plate is bounded by some combination of mid-ocean ridges (now often termed *spreading centers,* because they are locations where two adjacent plates move away from each other), trenches (where two plates converge), and *transform faults* (where two plates slide sideways past each other). A plate moves as a unit across Earth's surface. In 1967, Jason Morgan of Princeton University began the mathematical analysis of the plates' wanderings, by pointing out that these motions can best be described in the language of spherical geometry—a geometry in which distances become angles and motions become rotations about poles. A Frenchman, Xavier Le Pichon, used this new view to construct the first map of Earth's *tectonic plates.* There are 13 to 20 of these plates all told (depending on exactly how they are subdivided), but 7 of them are much larger than the rest. These are the Eurasian, African, North American, South American, Pacific, Indian-Australian, and Antarctic plates. By the late 1970s, the rates and directions of movement of all the plates had been calculated, and these calculations have been progressively integrated into a complete description of the movements of the lithosphere.

Because the plates' motions are calculated relative to those of their neighbors, the theory of plate tectonics makes no statement about which plates are "really" stationary with respect to Earth's interior and which are "really" moving. Nevertheless, we shall see in Chapter 9 that there is a set of relatively fixed markers in the mantle, called *hot spots,* that can be used to assign absolute speeds (or better, angular velocities) to the overlying plates.

It's worth mentioning that the view of the lithosphere offered by plate tectonics—that of completely rigid plates separated by hair-thin boundaries—is only an approximation of the truth. In reality, the boundaries between plates can be quite wide—as much as several hundred miles in some instances—and can be composed of many faults. And even the interiors of plates are not completely stable. As we'll see in Chapter 8, a large part of the American West, although deep within the North American Plate, is taking off on a journey of its own and is dramatically altering the landscape in the process. But

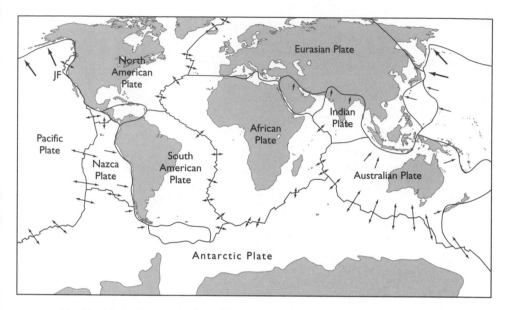

The Earth's major tectonic plates. The arrows indicate the direction and speed of convergence or divergence at plate boundaries. The longest arrows correspond to a speed of about 17 centimeters (6.7 inches) per year. JF indicates the small Juan de Fuca Plate, shown in greater detail in the next illustration.

still, the plate-tectonic picture of the Earth is central to the theme of this book, because the great majority of all earthquakes and volcanic eruptions occur on or near plate boundaries.

If we ask why the plates are in motion at all, the basic answer is heat. Heat from the Earth's interior has to find its way to the surface and ultimately be radiated away into space. This heat has three sources. First, the Earth started out hot—it was heated by the collision of the bodies from which it was formed, and a little of that initial heat still remains in Earth's interior. Second, new heat is constantly being generated by the decay of radioactive elements in Earth's rocks. And third, heat is generated by the slow crystallization of molten iron within Earth's core.

Some of this heat reaches the surface by direct conduction through the rocks of the mantle and crust. In addition, however, heat is brought to the surface by the actual movement of hot rock, especially by the upwelling of magma at the mid-ocean ridges. In essence, the tectonic plates form the upper parts of convection cells—looplike systems of material that circulate under the influence of gravity, transferring heat in the process. (Similar loops of upwelling and downwelling can be observed in thick soups or jams boiling on a stovetop.) As the newly formed crust moves away from the mid-ocean ridge, its heat is transferred to the overlying seawater. By the time the rock is reabsorbed at the deep-sea trenches, it is cold and dense and therefore ready to be drawn by gravity down through the warmer, lighter rock beneath.

When the descending (*subducting*) lithosphere penetrates into the deeper mantle, some mantle material is melted and rises toward the surface, by a process that we'll describe more fully in Chapter 2. This material in turn transfers its heat to the lower part of the crust and induces melting of portions of the lower crust. These molten materials rise like inverted teardrops and congeal in the upper crust, forming masses of granitic rock like those found in Yosemite National Park. The granitic rocks generated by this process of redistillation over billions of years are what form the continents. Too buoyant to be recycled into the mantle, the continental lithosphere is amazingly stable; some continental rocks date back nearly 4 billion years—25 times older than the oldest rocks on the deep ocean floor.

Despite the relative stability of the continental crust, the outlines of the continents have changed incessantly over Earth's lifetime. Wegener's supercontinent of Pangaea was only a temporary aggregation of the world's landmasses; before that, the continents had already aggregated and dispersed several times—about once every 600 million years.

When geologists in the 1960s and early 1970s estimated the speed and direction of motion of the tectonic plates from the historical record of the magnetic strips laid out on the ocean floor, they never imagined that the tectonic motions would one day actually be observed and measured in real time, for that would involve measuring the distance between, say, Paris and New York to an accuracy of an inch or less, and then repeating the measurement a year or two later with the same precision. Since the early 1980s, however, the U.S. Department of Defense's Global Positioning System (GPS) has given Earth scientists exactly such a capability. The use of this system, which depends on precise timing of radio signals arriving from several orbiting satellites, has confirmed that the long-term movements inferred from the paleomagnetic studies of rocks on the seafloor are continuing today. (The GPS technique is described in greater detail in the appendix.)

A Bang or a Whimper?

So, back to our story of earthquakes in the Pacific Northwest. By the 1970s, the general layout of the tectonic plates in that area had been mapped. A "mid-ocean ridge" called the Juan de Fuca Ridge runs roughly north-south, about 300 miles offshore—actually, it is interrupted by several offsets, known as transform faults, giving it a zigzag appearance. Closer to the mainland (about 60 to 100 miles offshore) the ocean floor begins its descent or subduction under the North American Plate. (There would be a trench here, were it not for the Columbia River, which has contributed enough sediment to choke it up.) The ocean floor between the "mid-ocean ridge" and the subduction zone is a relatively small tectonic plate named the Juan de Fuca Plate. (Juan de Fuca, by the way, was a sixteenth-century Greek explorer who may or may not have discovered the strait named in his honor that runs between Vancouver Island and the Olympic Peninsula.)

Attentive readers may already have smelled a rat: if the Juan de Fuca Ridge is a "mid-ocean ridge," why isn't it in the middle of the Pacific Ocean? Has it, for time immemorial, been churning out more seafloor on its western flank than on its eastern flank, thus causing the Pacific Ocean to expand asymmetrically? The answer is no. The reason that the ridge is so close to the North American continent is that the continent, moving steadily westward as the Atlantic Ocean continues to widen, has been overriding the Juan de Fuca Plate faster than it can be created. Thus, the Pacific Ocean is actually shrinking—at about the same rate as the Atlantic Ocean is expanding. In fact, the Juan de Fuca Plate (and some other small plates in the region of Central America) are mere remnants of an immense sector of seafloor posthumously named the "Farallon Plate," most of which has disappeared into the mantle under the Americas.

The sinking of the Juan de Fuca Plate takes place along the Cascadia subduction zone; the actual line of disappearance of the seafloor is the offshore "trench." The reason that the oceanic plate, rather than the North American plate, is being subducted is simply that oceanic rock is much denser than continental rock and therefore sinks more readily. In fact, at all converging plate boundaries where one plate is oceanic and the other continental, the continental plate is the preordained survivor.

During the 1980s, geologists established the rate at which the Juan de Fuca and North American plates are converging, using the paleo-

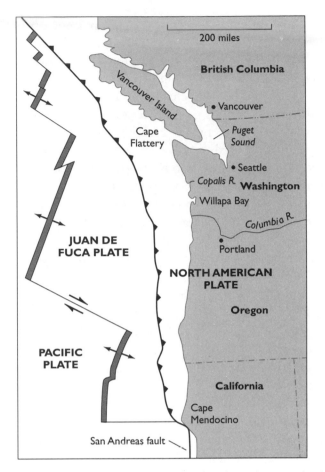

The Cascadia subduction zone. The zigzag line at left is the mid-ocean ridge (double line) offset by transform faults. The toothed line at center is the margin of the subduction zone, where the North American Plate is overriding the Juan de Fuca Plate.

magnetic technique. The rate is about 4 centimeters (1.6 inches) per year. That is the combined total of the eastward motion of the seafloor (3 centimeters or 1.2 inches per year) and the westward motion of the North American continent (about 1 centimeter or 0.4 inch per year).

For the first 10 years or so after the discovery of the Cascadia subduction zone, it remained a topic of purely academic interest. Because there had never been a great earthquake in coastal Washington or Oregon throughout recorded history, these regions were assigned the lowest or next-to-lowest seismic hazard ratings (a 1 or a 2), and so buildings could be constructed with little or no seismic reinforcement.

But in the early 1980s, construction began on a nuclear power station in the Coast Range south of Puget Sound, and the matter of seismic hazards had to be considered more carefully. In its license application, the Washington Public Power Supply System (WPPSS, pronounced "whoops!") took the position that, although there was indeed an off-shore subduction zone, this particular zone was unusual: subduction occurred by small increments or "creep," they argued, not by the large jumps that unleashed earthquakes.

The Nuclear Regulatory Commission asked the USGS to review the application, and the USGS dropped the application in the lap of Thomas Heaton, who then worked at the USGS's Pasadena office. "I looked at it," Heaton told us recently, "and any time somebody makes a claim like that you say to yourself, Why did they say that? It might be creep, or it might not, but I had to play devil's advocate. So I went over to Caltech and talked with Hiroo Kanamori, who was doing a lot of work on subduction zones with Seiji Uyeda of Japan."

According to Kanamori, the mechanism whereby an oceanic plate subducts seems to depend on the age on the plate. Near the Mariana Islands in the western Pacific, for example, the ocean floor subducting at the trench has traveled thousands of miles from its spreading center. It is now more than 100 million years old. At such an extreme age, the oceanic crust has become so cold and dense that it falls readily and steeply into the mantle, with little interaction between the subducting ocean floor and the overriding plate. At subduction zones like this, there are few if any large earthquakes but almost incessant tiny ones. Off the coast of southern Chile, on the other hand, the South American Plate has encroached so close to the mid-ocean ridge that the subducting ocean floor is barely 10 million years old—a youngster by geological standards. It is still relatively warm and buoyant and therefore not at all disposed to go under without a fight. In this type of subduction zone, according to Kanamori, the subduction process is jerky: the oceanic and continental margins press against each other with gradually increasing force, until the contact zone suddenly gives way, causing a giant earth-quake. In fact, when the subduction zone off southern Chile slipped in 1960, it unleashed the biggest earthquake ever recorded—a behemoth that not only devastated 600 miles of Chilean coastline but also sent tsunamis across the Pacific, drowning hundreds of people in Hawaii and Japan. A total of about 5700 people died. The earthquake earned a 9.5 on the *moment-magnitude* earthquake scale. (This scale, used throughout the book, is a measure of the total energy liberated during the earthquake. Each unit increase in moment magnitude represents an approximately thirtyfold increase in energy release. We describe the methods of measuring moment magnitude in the appendix.) Four years later, the second largest earthquake of the twentieth century, magnitude 9.2, rocked the Pacific coast of Alaska. It killed 131 people and caused extensive damage in Anchorage and elsewhere (see Color Plate 4). Again, it was caused by the subduction of relatively young ocean floor.

To Heaton, all the physical features of the Cascadia subduction zone, and most especially the young age of the subducting ocean floor, made it look like the Chilean type of subduction zone, not the Marianas type. The fact that no earthquake had been recorded on the Cascadia subduction zone was probably deceiving. For one thing, Europeans had inhabited the Pacific Northwest for only 200 years. Why couldn't it take more time than that to generate sufficient strain to force a rupture of the locked subduction zone? In addition, the seismic silence was too complete. In places like the Marianas, where the ocean is subducting by "creep," tiny but easily recordable earthquakes are going on all the time. In Cascadia, even these tiny earthquakes are absent. This suggested to Heaton that the subduction zone was not creeping but was totally locked up.

To test his hypothesis that the Cascadia subduction zone broke in infrequent, great earthquakes, Heaton searched for evidence that such an earthquake had occurred in the Pacific Northwest prior to the Europeans' arrival. With the help of a colleague, Parke Snavely, Jr., he found what he was looking for in a book entitled *The Indians of Cape Flattery,* written in 1868 by a collector of American Indian lore, Judge James Swan. Cape Flattery forms the northwestern tip of the Olympic Peninsula, at the entrance of the Juan de Fuca Strait. It is, in fact, almost an island, for only a 3-mile-wide neck of marshy ground, running between Neah Bay on the strait and Mukkah Bay on the Pacific, joins the cape to the rest of the peninsula. Swan wrote:

"A long time ago," said my informant, "but not at a very remote period, the water of the Pacific flowed through what is now the swamp and prairie between Waatch village and Neeah Bay, making an island of Cape Flattery. The water suddenly receded leaving Neeah Bay perfectly dry. It was four days reaching its lowest ebb, and then rose again without any waves or breakers, till it had submerged the Cape, and in fact the whole country, excepting the tops of the mountains. . . . The water on

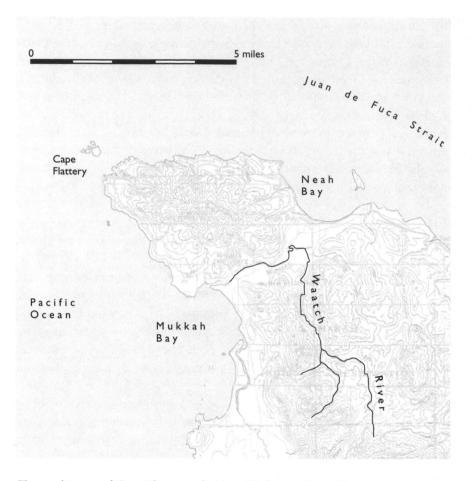

Topographic map of Cape Flattery and vicinity, Washington State. Contour intervals are 130 feet (40 meters).

its rise became very warm, and as it came up to the houses, those who had canoes put their effects into them, and floated off with the current, which set very strongly to the north. Some drifted one way, some another; and when the waters assumed their accustomed level, a portion of the tribe found themselves beyond Nootka [on the coast of Vancouver Island], where their descendants now reside ... Many canoes came down in the trees and were destroyed, and numerous lives were lost. The water was four days regaining its accustomed level."

Heaton and Snavely interpreted the account as a description of a tsunami, although some details seemed to have been distorted, such as the long duration of the tsunami as well as its great height—it would have to have been 1400 feet high to entirely inundate the cape. Although the account makes no mention of an earthquake, the most likely source of a very large tsunami was a slippage on the local Cascadia subduction zone.

Judge Swan had himself been curious enough to dig into the turf covering the marshy neck of land behind Cape Flattery. He found sand—evidence, he felt, that the Pacific Ocean did once completely surround the cape. But it wasn't until more than a century later that the real excavations began.

Digging Up Old Earthquakes

Brian Atwater moved to Seattle in 1985. He had previously been based at the USGS's western headquarters in Menlo Park, California. Before he left Menlo Park, Atwater heard Heaton give a lecture on the Cascadia subduction zone. Perhaps, he thought, he could take advantage of his move to test Heaton's ideas in the field—to see, in other words, whether there was any evidence of prehistoric giant earthquakes along the Northwest coast.

Atwater chose to begin his work at Cape Flattery because a tide gauge had been installed there many years earlier. From the sea-level measurements provided by the tide gauge, it was clear that that the Cape was rising out of the sea at a rate of about 3 millimeters per year. If this rate of uplift continued over long periods, shoreline features on Cape Flattery should come to lie much higher than corresponding features farther east, say on Puget Sound, where uplift was not taking place. But, in fact, there was no evidence of such cumulative differences in level. Could it be, Atwater thought, that the periods of uplift were interrupted by sudden drops—drops that were associated with giant earthquakes?

Atwater started digging where Judge Swan had dug. A little tidal stream, the Waatch "river," meanders across the low-lying terrain behind Cape Flattery. The stream has partially cut through the underlying layers, thus doing most of Atwater's work for him. By shoveling away at the banks of the stream, he got a good view of the soil layers several feet below the surface.

Indeed, there was sand and mud right below the surface, and it extended for several feet downward. But then, very sharply, the sand gave way to a layer of pure black peat. Digging down even farther, the peat gradually became mixed with sand and mud again.

This kind of layering is a chronological record: the lowest layers are the oldest and the highest layers the youngest. The fact that a layer of sand was followed by a layer of peat, then by another layer of sand, and finally by the humus of the topsoil suggested that this particular spot of ground had been below the high-water mark, then above it, then below it, and finally above it again. Most intriguingly, the sharpness of the peat-to-sand border suggested that the change from above to below the high-water mark had taken place quite suddenly. This was Atwater's first piece of evidence that a sudden drop in the shoreline had indeed occurred at least once in the past.

Atwater knew that subduction earthquakes are commonly accompanied by changes in the height of the coastal land. Typically, the gradual accumulation of stresses at the locked subduction zone deforms the continental plate as if it were a giant foam-rubber mattress, causing the land surface to bow gradually upward in one place and to bend downward elsewhere. Of course, rock cannot deform as much as foam rubber. But when compressed over scores of miles, a plate can shorten by tens of feet horizontally and expand correspondingly in thickness, raising the land surface. When the strain is released, on the other hand, the plate lengthens and thins, causing the surface to drop. Thus, land that is near the sea may gradually rise above the water level and then suddenly drop below it.

If this was the explanation for the buried peat layer at Cape Flattery, the layer should exist at other places along the Northwest coast. So Atwater, as well as other researchers who were excited by his discovery, went up and down the coast digging for signs of ancient earthquakes— a relatively new field of research that has come to be called *paleoseismology.* They found similar layers almost wherever they looked. At several sites, indeed, there was not one but several layers of buried peat, separated by layers of mud (see Color Plate 5).

The prize find was made at Willapa Bay in southern Washington, a few miles north of the mouth of the Columbia River. Here Atwater found no fewer than six peat layers—the record, apparently, of six successive cycles of subsidence and uplift. Furthermore, at least three of the peat layers were capped with a thin layer of pure sand that, as Atwater dug farther inland, became gradually thinner and finer-grained. These layers, Atwater surmised, were deposited by tsunamis that swept across the low-lying areas immediately after the earthquakes (see Color Plate 6).

Atwater wanted to know the dates of the layers, so that he could estimate when and how frequently the subduction zone had broken. For this purpose, he took fragments of wood and other plant material from the peat and sent them to a laboratory for radiocarbon dating, a technique that estimates a sample's age by measuring how much of the radioactive isotope carbon-14 has decayed to nitrogen-14. (We describe this technique in greater detail in the appendix.)

The results from Atwater's samples indicated that the most recent subsidence took place about 300 years ago, with an uncertainty of a couple of decades. Dates for the earlier events were much less certain. Still, Atwater's evidence spoke for at least six subsidence events, and possibly several more, within the past 4000 years.

Besides the buried peat layers, Atwater found other kinds of evidence bearing on past subduction earthquakes. At the estuary of the Copalis River, about 90 miles south of Cape Flattery, there stands a forlorn, dead forest of red cedars, their roots killed by the salty intertidal water in which they are sitting. Evidently, the trees grew when the ground was higher and were killed when the ground suddenly sank below the high-water mark. Study of the rings in these standing trees could not give the exact year of the subsidence, because the outer rings had rotted away. But the outermost surviving rings are from about the year 1680, suggesting that the trees died a few years after that. In other words, they probably died in the same subsidence event that buried the uppermost peat layer at Willapa Bay, which was carbon-dated at about 300 years old.

In the same general area of the Copalis River, Atwater found localized masses of sand that appeared to have been blown up through the topsoil from deeper layers as if by the action of miniature volcanoes. These "sand-blows" typically occur in areas where wet sand lies only a few feet below the surface, when the terrain is subjected to severe shaking. During the shaking, the hydrostatic pressure (water pressure) in the sandy layer rises to the point that the sand grains separate from each other and the whole layer turns into a slurrylike quicksand—a process appropriately referred to as *liquefaction*. The liquefied mixture of sand and water then erupts through the overlying layers onto the surface (see Color Plate 7). This was further evidence that the drop in the land that took place about 300 years ago was probably not a gradual subsidence

Red cedar trees killed by subsidence during a great earthquake. Copalis River estuary, Washington State. The photo was taken at low tide; at high tide the water almost reaches the top of the bank. (Brian Atwater, USGS)

but a sudden fall, accompanied by the kind of severe shaking that characterizes a very large earthquake.

Perhaps the most detailed record of past giant earthquakes in the region has been obtained not from the land but from the deep ocean floor just beyond the continental slope. In the late 1960s, a group of Oregon State University students, under the supervision of L. D. Kulm, drilled into the ocean floor and recovered long cores for study in the laboratory. The cores consisted mostly of mud—the typical ocean-floor material that accumulates from sediment raining gently out of the overlying seawater. But within the mud were layers of a sandy material. The students counted 12 sandy layers before they came, near the bottom of the cores, to a layer of volcanic ash. The students identified the ash layer as resulting from the eruption of Mount Mazama, a volcano in Oregon that blew its top about 7700 years ago (see Chapter 2).

In the mid-1980s John Adams, a geologist with the Geological Survey of Canada, reexamined the students' data. He hypothesized that each sandy layer resulted from violent shaking of the slope of the continental shelf during a subduction earthquake. As a result of the shaking, sand was stirred up from the shelf and poured down the slope, coming to rest on top of the deep-ocean mud. As time passed, more sedimentary mud built up, until another earthquake delivered another flow of sand, and so on. Thus, if Adams's reasoning is correct, there have been at least 12 giant earthquakes on the Cascadia subduction zone over the past 7700 years.

Thus, both comparative studies and observations in the field tell us that the Cascadia subduction zone does produce giant earthquakes and that the most recent earthquake happened about 300 years ago, give or take a few decades. A curious circumstance has allowed this most recent earthquake to be dated to the precise year, date, and even time of day. In 1994, Kenji Satake, a seismologist with the Geological Survey of Japan, attended a paleoseismology workshop in California at which the history of subduction earthquakes in the Pacific Northwest was discussed. He knew that large subduction earthquakes can cause tsunamis that cross the Pacific Ocean. Therefore, when he returned home, Satake and his colleagues combed through old tsunami records covering the period around 300 years ago. They found several documents that mentioned a tsunami that came ashore in Honshu in January 1700. By carefully eliminating all other potential sources, they concluded that this tsunami was probably caused by a rupture of the Cascadia subduction zone. Knowing the speed at which tsunamis travel across the ocean, Satake and his colleagues were able to fix the time of the earthquake at about nine in the evening, local time, on January 26, 1700. And on the basis of the size of the tsunami, they were able to estimate the size of the earthquake: it had a magnitude of about 9.0—in the same size range, in other words, as the giant Chilean earthquake of 1960 and the Alaskan earthquake of 1964.

A final piece of evidence clinches the story. Having heard of Satake's work, Atwater went back to his dead red cedar trees on the Copalis River. Although the outermost rings on the trees' trunks had weathered away, the rings in the some of the submerged roots were fully preserved. Dating root rings is notoriously difficult, but by tracing individual rings in the roots up into the trees' trunk, where they could be dated, Atwater was able to establish the year of the trees' final growing season: it was 1699.

The Prospects

For residents of the Pacific Northwest, a decade of research has brought dramatic and unsettling news. Extremely large earthquakes, often accompanied by tsunamis, have occurred in the region—and will occur again.

Tom Heaton has summed up the science: "I think this particular story of the Cascadia subduction zone is probably one of the more amazing threads of detective work in our business. I mean, there are no historic records of a large earthquake up there. And from the beginnings of understanding seafloor spreading, people then understood that this has got to be a subduction zone here, there's got to be convergence across the area. And then from analogies with other places there were suggestions that it may well move in great earthquakes, leading people to look for tsunamis and subsidences in the geologic record, and then actually finding historic accounts of these things in some other place. Things hold together."

But the science is also painfully inadequate. The burning practical questions are: When and where will the next earthquake occur in the Pacific Northwest? What areas will it affect? How strong will it be? The answers to these questions involve much guesswork, beyond the frontier of the known science.

Because the latest giant subduction earthquake took place 300 years ago, and because these events seem to repeat at intervals ranging from 300 to as much as 900 years, it seems fair to say that we are now entering a broad time period within which the next earthquake will occur. But this period of concern may be several hundred years long. There is no reason to assert that the next earthquake is overdue or even due in our lifetime. Quite possibly, Northwesterners will have to live with the risk of a giant earthquake for many generations before it actually comes to pass. The only reason to shorten the prediction would be if further fieldwork shows that earthquakes occurred more frequently in the past than our current estimates suggest or if warning signs of an impending earthquake are detected.

The size of the coming earthquake is also uncertain—although the choice is basically between a series of "great" quakes and a single "giant" one, depending on whether the Cascadia subduction breaks as a series of separate segments or as a single slip event along the entire 750-mile front. If the subduction zone breaks as separate segments, each earthquake will probably have a magnitude of about 8. Geologist George Carver, of California State University, Humboldt, has called this the "decade of terror" scenario. Each event might well release at least 30 times more seismic energy than the 1995 earthquake that devastated Kobe, Japan, over a region several times larger.

If the entire subduction zone breaks at once, the resulting earthquake would probably have a magnitude of about 9—about a *thousand-
fold* more energetic than the Kobe quake, putting it about on a par with
the 1960 Chilean earthquake. The present evidence is too fragmentary
to be sure which of these two scenarios will happen. It does seem probable, however, that at least the most recent of the previous earthquakes
on the Cascadia subduction zone involved a single rupture of the entire
zone.

Such a giant earthquake would shake the entire coastline with
extreme violence for several minutes, and a tsunami might well follow
the shaking, inundating low-lying areas. The shaking would be somewhat less violent in such cities as Vancouver, Seattle, and Portland,
which lie inland from the coast. These cities have high-rises of up to 76
stories. "They might ride out a subduction earthquake," Heaton commented, "but I'd rather not be there for the experiment!"

Many other regions of the Pacific Rim are threatened by great subduction earthquakes. These include the entire west coast of South
America, New Zealand, Sumatra, Japan, the Aleutian Island chain, and
Alaska. Tokyo is particularly at risk: it lies in a geologically complex
region where four plates—the North American Plate, the Eurasian
Plate, the Pacific Plate, and the small Philippine Sea Plate—grind
against one another. The margin of the subduction zone at which the
North American Plate is overriding the Philippine Sea Plate comes on
land between Tokyo and Mount Fuji. There are historical records,
extending back 13 centuries, of giant earthquakes on this and nearby
subduction zones.

As mentioned earlier, there are some subduction zones where most
of the motion between the converging plates occurs by gradual "aseismic" creep rather than by infrequent large ruptures. The subduction
zone along the western edge of the Pacific Plate, in the region of the
Marianas Island chain, is a good example. But even here, seismic hazards do exist: Guam, at the southern end of the chain, was recently
struck by a magnitude 8.0 earthquake.

What we have described in this chapter is the typical behavior of
converging plates where one plate carries oceanic crust and the other
carries continental crust: because the continental crust is lighter, it over-
rides the oceanic plate. But there are a few regions in the world where
two plates, both carrying continental crust, are converging. The most
striking example is the border of the Indian and Eurasian plates: the

Indian Plate is moving northward, relative to the Eurasian Plate, at about 50 millimeters (2 inches) per year. Because both plates carry relatively buoyant crust, neither wants to "go under"; the result is the piling up of the Himalayan Range and the Tibetan Plateau. A set of faults running east-west along the southern foothills of the Himalayas has broken in four great earthquakes over the last century—in 1897, 1905, 1934, and 1950. Each had a magnitude of about 8.5. Collectively, these ruptures have broken much of the fault, but a 400-mile stretch, from northeast of Delhi to west of Kathmandu, remains unbroken in historic times. This "seismic gap" will probably rupture at some point in the future, causing another very destructive earthquake.

Earthquakes are not the only geological hazards created by the process of subduction. In the next chapter, we follow the descending Juan de Fuca Plate deep into the mantle, where it is slowly reheated. This reheating is the ultimate cause of the Pacific Northwest's second and more visible geological menace—the volcanoes of Cascadia.

2

BLASTS FROM THE PAST: MOUNT ST. HELENS AND HER SLEEPING SISTERS

"*Vancouver . . . Vancouver . . . this is it!*"

These were the last words that 31-year-old volcanologist David Johnston yelled into his radio as he alerted USGS headquarters in Vancouver, Washington, to what was taking place in front of him. It was 8:32 on the morning of May 18, 1980. Johnston was standing on a high ridge 6 miles from the summit of Mount St. Helens. Moments earlier, he had seen the entire north side of the mountain begin to fall away. As the immense mass of rock and ice slid downward, the mountain's molten core erupted in two terrific explosions, one directed upward, the other northward, toward Johnston's position. It took less than two minutes for the blast cloud to reach him. Johnston, and at least 56 other people in the surrounding area, died from the blast, from the heat, or from the choking ash that rained down so thickly as to turn day into blackest night.

Volcanology is a dangerous trade—especially for the tiny band of researchers who study ongoing eruptions at arm's length. About every two years, on average, a volcanologist dies in the line of duty. As recently as January 1993, six scientists were killed by a small eruption in the crater of the Galeras volcano in Colombia as they were installing monitoring equipment. The lone survivor of the party was its leader, Stanley Williams of Arizona State University, who suffered two broken legs and a fractured skull. Undeterred, Williams was soon back in the business: in the spring of 1996 he sampled gases being emitted by Popocatepetl, the volcano that towers menacingly over Mexico City.

Volcanologists take these risks not because they are foolhardy adventurers but because of their fascination with their subject and because of the great potential for human benefit that their work offers. Unlike earthquakes, most volcanic eruptions are visually spectacular and are preceded by signs that, properly interpreted, can help avert enormous disasters. If it had not been for the warnings issued by scientists, and the actions taken by government agencies in consequence of these warnings, the precursory events at Mount St. Helens would have drawn thousands, perhaps tens of thousands, of visitors to the flanks of the volcano. Yet even at Mount St. Helens, the timing and course of the main eruption were not forecast precisely, and that is part of the reason that Johnston and many of the others died. Despite the best efforts of volcanologists, eruptions still cause great loss of life around the world. In 1985, for example, the eruption of the Nevado del Ruiz volcano in Colombia killed 25,000 people. Thus, there is every reason to intensify research into volcanoes, to monitor volcanoes more closely, and to improve the means by which scientific information is translated into practical measures to safeguard the population.

Three main lines of research help define the hazards posed by volcanoes. First, there is the study of the underlying geological processes that give rise to volcanoes. Second, there is the study of the history of previous eruptions at any particular volcano—a history that suggests the range of scenarios that might be followed in a future eruption. Last comes the most dangerous part—the study of the precursory events as a dormant volcano comes back to life.

The Ring of Fire

Stopped in the street and asked to explain how volcanoes work, the average citizen might give an answer something like this: a volcano contains a tube that runs down to a region of the Earth so deep and so hot that the rock is molten; once in a while, this molten rock forces its way up and erupts as lava. If pressed to explain why volcanoes are so common in certain places—fully three-quarters of all active volcanoes above sea level surround the Pacific Ocean, and many of the remainder are in or near the Mediterranean Sea—our interviewee might propose that in volcano country the Earth's crust is especially thin; riddled with cracks; or composed of weak, easily penetrated rock.

For a long time, this was also the scientific theory of volcanoes. But, as we mentioned in the previous chapter, the study of seismic waves has proved that Earth's mantle is generally solid. It is kept solid, despite its high temperature, by the enormous pressure exerted by the tens and hundreds of miles of overlying rock. Only in a few spots are mantle rocks in a partially molten state, and these spots are usually underneath volcanoes. So the question of why volcanoes crop up in certain places becomes the question of why, in these places, the mantle rocks are partially molten.

Before the development of the theory of plate tectonics, geologists still attempted to answer this question in terms of local peculiarities of the Earth's crust. Howel Williams, the leading expert in the "pre-plate-tectonic" era on the volcanoes of Cascadia, expressed a commonly held view when he stated that mantle rocks melted when the pressure on them was partially relieved by cracking of the overlying crust. These cracks, he and others speculated, occurred because Earth's surface was adjusting itself to a gradually cooling (and therefore shrinking) core. The Pacific Rim, he supposed, was characterized by many such cracks; therefore, many volcanoes are found there.

Nowadays, the volcanoes of the Ring of Fire are understood in the context of the subduction of seafloor at plate boundaries. In essence, the eruptions of these volcanoes are the parting shots fired by the oceanic lithosphere as it descends obliquely into the mantle to be recycled.

In the course of its descent, a cool oceanic plate is reheated by the surrounding mantle. If the descending slab consisted purely of the original basalt that emerged at the mid-ocean ridges millions of years ago, the rewarming would have no dramatic consequences. But in reality,

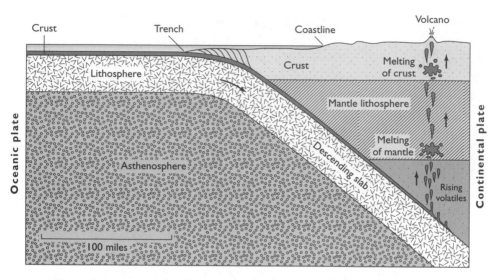

How volcanoes are produced by a subducting slab.

other materials have been added. Most notably, large quantities of water entered the basalt during the long period that it formed the ocean floor. In addition, waterlogged ocean sediments and fragments from the continental margins are dragged down with the descending slab. By the time the slab is about 80 miles below the surface, the rock has heated to over 1000°C. Because of the great pressures at this depth, this temperature is not high enough to melt the rock, but the heat does drive off the water and other volatile compounds, which rise into the overlying mantle. This mantle material is as hot as or hotter than the slab descending beneath it, and being nearer the surface it is under less pressure. Thus, as the rising volatiles dissolve into this hot rock, the rock's melting point is lowered enough for melting to occur locally. The resulting blobs of molten rock, or *magma,* percolate upward into the lower crust, where they in turn cause part of the crust to melt. The magma produced by this secondary melting aggregates in large chambers a few miles below the surface. The magma can sit in these chambers for many thousands of years.

In the Pacific Northwest (as also in Central and South America), the angle of descent of the ocean slab beneath the continent is rather gradual, so that it has traveled about 200 miles eastward from the subduction trench, and thus about 150 miles inland from the coastline, before it is deep enough and hot enough to release its burden of volatiles. Thus, the magma chambers are formed at about this distance from the coast, and this is why the Cascadian volcanoes themselves are so far inland. In much of the western Pacific, by contrast, the relatively cold oceanic slab descends more steeply into the mantle. Therefore, the resulting volcanoes tend to form arcs of islands closer to the trenches.

Much of the molten rock within a magma chamber solidifies in place and goes no farther. These solidified parts consist of the "granites" or "gabbros" that are familiar to us as building materials. But sometimes, a portion of the magma is able to make its way through conduits in the overlying rock and reach the surface. A volcanic eruption is the result.

All of the large volcanoes of Cascadia, from Mount Lassen in northern California to Mount Garibaldi in British Columbia, are *stratovolcanoes* ("layered volcanoes"). Many of the world's best known volcanoes, such as Mount Vesuvius in Italy, Mount Fuji in Japan, and Krakatau in Indonesia, are of this type. In its prime, a typical stratovolcano has a high conical shape and is built of layers that slant steeply downward from the peak. The layers are composed of solidified lava flows interspersed with layers of ash and rock fragments that fell out of the air or were carried downhill in roiling masses of hot gas. Stratovolcanoes are usually fed by magmas of varying compositions, sometimes low in silica (basalt), but more often high in silica (dacite, rhyodacite, rhyolite) or of intermediate composition (andesite). The composition varies either because the magma in the underlying magma chamber has separated out into different chemical fractions or because the volcano vents the contents of two or more magma chambers that contain different kinds of magma.

The Incredible Importance of a Tiny Molecule—Silica

Silica (silicon dioxide) is the main constituent of all but some very unusual magmas. The exact proportion of silica in the magma, however, is crucial to a volcano's behavior. The amount of silica affects whether or not a volcano is highly explosive and whether the volcano's shape is bulky and gently sloping, like the volcanoes of Hawaii, or steep and rugged, like Mount Fuji and Mount St. Helens.

The influence of the proportion of silica is related to a property of silica molecules in molten rock: their ability to cross-link with other silica molecules, forming a tough, epoxylike network. When the silica content is low, as in basalt, the magma is relatively thin and runny.

When this kind of magma rises in the volcano's conduit, and the pressure exerted by the overlying rock decreases, water and other volatiles dissolved in the magma come out of solution and form bubbles. As the magma continues to rise, the bubbles expand and collect together. Thus, by the time the magma reaches the surface, the gases originally dissolved in the magma have largely separated from it. Therefore, there is no great head of pressure within the magma that could give rise to an enormous explosive eruption. In general, the danger from basaltic eruptions is not that the mountain will blow its top but that the lava issuing from the vent will flow for many miles downhill, destroying habitations, roads, and anything else in its path. This is the standard kind of eruption produced by Hawaii's basaltic *shield volcanoes,* to be discussed in Chapter 12.

If the magma has a high silica content, the extensive cross-linking between silica molecules makes the molten rock much more viscous. During the magma's rise toward the surface, bubbles form in the same way as they do with basaltic magma. But the stiff matrix of the magma—which becomes rapidly stiffer as the water is removed from it—prevents the bubbles from enlarging. Therefore, as more and more steam and gas enter the bubbles, the pressure inside them rises until it approaches the pressure exerted by the overlying rock—the so-called *overburden pressure.* At this point, any sudden drop in the overburden pressure—caused by a sudden fracturing of the rock or by a landslide on the volcanic edifice, for example—allows the bubbles to rip the frothy, silica-rich matrix into millions of cobble- and pebble-sized chunks of the Styrofoam-like material that we know as *pumice*—a material often so light that it floats in water.

It is this transition that is the main powerhouse behind the spectacular vertical eruptions for which many stratovolcanoes are famous. Such eruptions, which can carry pumice, ash, and hot gases miles into the stratosphere, take the form of *Plinian columns,* named in memory of the Roman statesman Pliny the Younger, who described such an eruption of Mount Vesuvius in 79 C.E. During the eruption of a Plinian column, the volcano behaves like a rocket engine stood on its head: the "fuel" is the magma rising through the volcano's conduits, which expands explosively at or just below the surface and is propelled skyward at hundreds of feet per second. On account of its great heat, the resulting mixture of air, volcanic gases and shredded pumice rises buoyantly into the atmosphere, sometimes to an altitude of nearly

100,000 feet. The column expands sideways as it rises, so that its appearance is roughly that of an upside-down Christmas tree. If there are winds aloft, the column of debris is blown downwind, falling back to Earth in an elliptical band that extends tens to hundreds of miles away from the volcano.

Despite the rocket analogy, it should be emphasized that combustion plays little role in volcanic eruptions. Somewhat confusing in this context is the term ash, which is used to describe the most finely pulverized material that may be carried for great distances downwind from the volcano. This term is a bit of a misnomer: the "ash" is not the product of burning but of the explosive disintegration of gas-charged magma into a fine powder.

Not all the magma within stratovolcanoes is ejected via Plinian columns. Some magma may be gas-poor, as is often the case with basaltic magma. Gas-poor magma pours out of vents and flows as molten lava down the flanks of the volcano. Often, even medium- or high-silica magma is not vapor-rich enough to cause a Plinian-style eruption. In such cases, the high-silica magma may also emerge as lava and, because of its high viscosity, harden in place near the volcano's vent. These lava flows are often easily recognizable; looking at Mount Lassen from the west, for example, one can see a stubby dark lava flow, made of dacite, that clings to the very summit of the mountain. The buildup of such flows over time tends to make the flanks of stratovolcanoes precariously steep. Thus, it is not uncommon for large parts of the volcanic edifice to collapse in catastrophic avalanches.

Why Mount Mazama Is Not on the Map

A placid, 5-mile-wide lake, reflecting a ring of 2000-foot cliffs in its deep blue waters, is not everyone's idea of a volcano. A casual visitor to Crater Lake in southern Oregon might imagine that it was created by some other means—perhaps by the impact of a meteorite. But here, more than 7000 years ago, a 12,000-foot mountain opened its volcanic mouth and discharged the contents of its enormous magma chamber. About 12 cubic miles of molten rock jetted skyward during that eruption, to fall as ash over an area of about 500,000 square miles—an area that included parts of what are now eight states and three Canadian provinces. As it disgorged this stupendous load of rock, the summit of the mountain

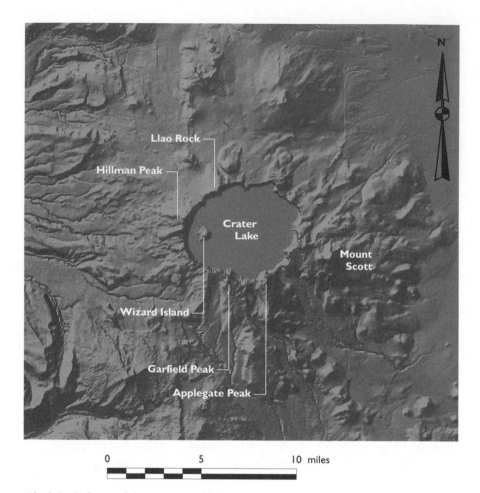

Shaded-relief map of the remnants of Mount Mazama, Oregon. Crater Lake occupies the caldera formed during the climactic eruption and summit collapse 7700 years ago.

sank into the emptying magma chamber, leaving a 4000-foot-deep, 5-mile-wide hole in the ground—a *caldera*. Half-filled with water, the caldera now lies in repose, a mute witness to Nature's past violence and a warning of what someday, somewhere, will happen again.

Of all the astonishing facts about Mount Mazama, as it has been named posthumously, one stands out as a purely human achievement: the local Native American population, the Klamath Indians or their predecessors, not only witnessed the mountain's cataclysmic destruction but also were able to maintain an oral history of the event over the thousands of years that followed it, retelling it to Europeans in the nineteenth century.

We know that Native Americans were there at the time of the climactic eruption because human remains (both bones and cultural artifacts) have been found in the ash. But the legend of the origin of Crater Lake was first written down by a soldier at Fort Klamath, William Colvig, who in 1865 heard it independently from several men of the local Klamath tribe.

At the site of Crater Lake, according to the legend, there was originally a high mountain. An opening at the mountain's summit led down to the dwelling of the Chief of the Below World. This Chief fell in love with an Indian maiden, but she refused to return with him to his subterranean home. In vengeance he stood atop the mountain, shaking the Earth. Fire spewed from his mouth, so that red-hot rock and ashes fell from the sky. The people fled in terror. Medicine men declared that only human sacrifice could turn away his wrath, but none of the young men would volunteer themselves. Therefore, during the night, two of the medicine men themselves climbed Mazama and leaped into the flaming vent. The Chief of the Above World, watching from Mount Shasta, approved of the sacrifice. He drove the Chief of the Below World into his home, collapsing the mountain on top of him. In the morning the high mountain was gone. For many years thereafter, rain fell and filled the great hole that had been made.

Since the summit of Mount Mazama now lies thousands of feet below the surface of Crater Lake, one might wonder how modern geologists could add anything to the Indian account. Certainly, we will never know in detail what the mountain looked like or the precise sequence of events that slowly built it up and suddenly tore it down. But still, there is an abundance of clues that have allowed the main outline of the mountain's history to be discerned. Prominent among the scientists who have interpreted these clues are two men: the late Howel Williams of the University of California at Berkeley, who published studies of the volcano in the 1940s, and Charles Bacon, of the USGS at Menlo Park, who extended and reinterpreted Williams' findings in the 1980s. Together, these men's work at Mount Mazama has provided crucial insights into a stratovolcano's life history.

Perhaps the most obvious clue is the "Mazama ash" mentioned previously. This layer is found throughout the Pacific Northwest. It is thickest (up to several hundred feet) and coarsest in the immediate neighborhood of the caldera, but even 100 miles downwind (to the northeast) the layer is still a foot thick. Radiocarbon dating, carried out on plant material found in or under the ash layer, has allowed the eruption to be dated at about 7700 years ago. So extensive and heavy was the ashfall that many of the Native Americans in the area must have died, either as a direct consequence of the eruption or because their means of sustenance were destroyed.

Some of the prior history of Mazama is preserved in the many lavas, craters, cones, and other signs of volcanic action that ring the caldera. Just to the east of the lake, for example, is the highest remaining volcanic cone in the area, the 8926-foot-high Mount Scott. This is the oldest extant portion of the Mazama complex—a volcano that started growing about 420,000 years ago. The Mount Scott cone became extinct when the focus of volcanic activity shifted a few miles westward. Mazama proper consisted of many individual volcanic cones over the course of its long lifetime. The highest point on the rim of the present caldera is Hillman Peak, situated on the west side of the lake. It is the outer half of a volcanic cone that formed about 70,000 years ago: its inner half fell into the caldera during the climactic eruption 7700 years ago. On the south rim, Garfield and Applegate peaks are the remaining slopes of an older cone whose vent was higher on the mountain—possibly at its summit, 4000 feet or so above the highest point on the present caldera. Hundreds of lava flows that originated in now-vanished cones are clearly visible in cross section in the walls of the caldera. These include Redcloud Cliff on the northeast side of the lake, a lava flow that was vented about 30,000 years ago, and the especially obvious black summit of Llao Rock, on the northwest side of the lake, which emerged as a lava flow no more than a century or so before the climactic eruption (see Color Plate 8).

The magma vented by Mazama early in its history varied in composition but was made predominantly of andesite, the medium-silica rock that, when molten, usually erupts without great explosivity. About 50,000 years ago eruptions began to turn out highly viscous, silica-rich rhyodacitic magma. After the Redcloud eruption, Mazama seems to have lain quiet for over 20,000 years. A geologist, if any had been around at the time, might have been tempted to declare the volcano extinct. But in the geological scheme of things, 20,000 years of dormancy is often a mere afternoon nap—time enough to lull the imprudent into believing a giant has died. The Llao Rock lava flow, and the main eruption that followed one or two hundred years later, proved that Mazama was merely hibernating.

The main sequence of events during the climactic eruption can also be inferred from surviving geological clues. On the remaining outer flanks of the volcano, and exposed in the walls of the caldera, are layers of deposits from the eruption. The deepest layer, which is up to 65 feet thick, consists of relatively large fragments of pumice. The fragments lie separate from one another in cobbly beds, which means that they were thrown high enough in the air to cool as they descended. Therefore, the eruption in its initial stages sustained a Plinian column, as we can also conclude from the wide distribution of the Mazama ash.

Above the deepest layer on the northern and eastern flanks of Mount Mazama is a second layer of pumice, up to 30 feet thick, in which the individual lumps are fused together. This "welded tuff" must have formed from pumice that was still hot when it reached its final resting place. Bacon interpreted this to mean that the pumice in this deposit was not raised high into the air in a Plinian column. Probably this was because the vent became wider during the course of the eruption, eventually allowing so much magma to reach the surface that it could not be lifted to a great height by expanding gas and air. Instead, it exploded radially outward from the volcano, as a *pyroclastic flow.*

Pyroclastic flows are among the most remarkable and dangerous of the phenomena associated with volcanoes. They consist of masses of pumice, ash, and hot gas that behave like fluids. They emerge either directly from the volcanic vents or by falling out of a low eruption column. In either case, they descend the slopes of the volcano at speeds of 90 miles per hour or more. They do not slow down when they reach level ground but are carried forward by the momentum of their enormous mass and speed, totally destroying all vegetation and animal life in their path and sometimes even ripping up the soil itself to a depth of many feet. Running to high ground may be of little use in escaping a pyroclastic flow, as they are sufficiently buoyant and swift to rise over large obstacles. During the climactic eruption of Mazama, for example, one pyroclastic flow overtopped the neighboring peak of Mount Scott.

The reason for the speed and buoyancy of pyroclastic flows is the constant resupply of hot gas from three sources: the degassing of the still-incandescent pumice within the flow, the heating of cold air entrained within the flow's leading edge, and the vaporization of trees and plants. As the flow loses heat and vapor, it begins to solidify and eventually collapses to the ground. But this may be tens of miles from the volcano.

The welded-tuff deposit was produced while the volcano was still erupting from a single vent. We know this because it contains fragments of rock, interspersed among the pumice, that must have been scoured from the walls of the vent. Because these fragments are similar in composition wherever the welded tuff is located, there could have been only a single source.

The uppermost layer of deposits is quite different from the two underlying layers. This layer, which is up to 65 feet thick near the rim and increases to 300 feet thick or more in some depressions adjacent to the volcano, consists of deposits left by pyroclastic flows that took place during the collapse of the caldera. When Bacon examined the composition of the pyroclastic flow deposits at Crater Lake, he found that the fragments of older volcanic rock contained within the deposits varied markedly from site to site around the caldera rim. From this observation, he concluded that these flows came from many different vents that were scattered in a roughly ring-shaped array around the volcano.

What caused the change from a single-vent discharge to a multiple, "ring-vent" discharge late in the course of the eruption? According to Bacon, the change marked the beginning of the collapse of the mountain into the emptying magma chamber. This process can be envisioned as the descent of a weighty piston into an oil-filled cylinder. To permit the piston to descend, fractures first had to come into being, separating a central core of the mountain, including its summit, from the surrounding rock. The core then began to descend (either as a unit or as many fragmented blocks), forcing magma up between the core and the surrounding rock and ejecting it from a ring-shaped array of vents. The eruption became especially rapid and violent at this point, so that the eruption column rose much higher than in the previous phase. Thus, when ash and pumice fell out of the column, they accelerated to great speeds. The resulting pyroclastic flows had the momentum to surmount virtually any obstacles in their path.

After the pistonlike summit region had subsided by several thousand feet, the eruption came to an end. It did not cease because the magma chamber was completely empty but because the magma remaining within it was so viscous that the entrapped gases were not released, even when the magma was decompressed. It is also possible that the remaining magma had a lower water content to start with, so that the vapor pressure inside it never approached the overburden pressure. In either case, the magma remained stagnant within the vents, cooled, and sealed them off, and the eruption was over. The entire eruption probably lasted no more than a few days.

Exactly how far Mount Mazama's summit descended during the formation of the caldera has not been determined, because the evidence now lies far beneath the floor of the lake. But from studies of ancient calderas whose interiors have been opened up by erosion, it would seem likely that "caldera fill"—a chaotic mixture of summit rock, landslide material from the caldera walls, pumice, and ash—reaches down as much as 15,000 feet below Mount Mazama's original summit. The magma that once filled this huge cavity is now spread out as ash over half a million square miles of western North America.

There have been later, minor eruptions within the caldera. Very shortly after the collapse, an andesite dome formed in the center of the caldera; it is now submerged beneath the lake. Within a few hundred years, two larger cones were formed; the summit of one of them still rises above the lake surface as Wizard Island. The most recent volcanic event at Crater Lake was the extrusion of a small rhyodacite dome on the northeast flank of Wizard Island, about 5000 years ago.

Crater Lake is now quiet. It may lie dormant for ages—so long, perhaps, that erosion will cut through the caldera's walls and let the whole lake drain away, leaving a rock rimmed meadowland. Or maybe Mazama will rise again, climbing eruption by eruption until it has restored itself to the stature of its ancient edifice.

Mount St. Helens—The Prelude

The recent eruption of Mount St. Helens was a low-key performance, certainly, when compared with Mazama's final cataclysm. But still, it was the largest volcanic eruption in the contiguous United States in historical times, and it was the first since the lesser eruptions of Mount Lassen between 1914 and 1917. It was the most closely monitored eruption in history and yet, when the mountain finally blew its top on May 18, 1980, it caught everyone by surprise.

Since the 1950s, Mount St. Helens had been the object of intensive study by USGS scientists, especially Donal Mullineaux and Dwight "Rocky" Crandell. These scientists were motivated to focus on Mount

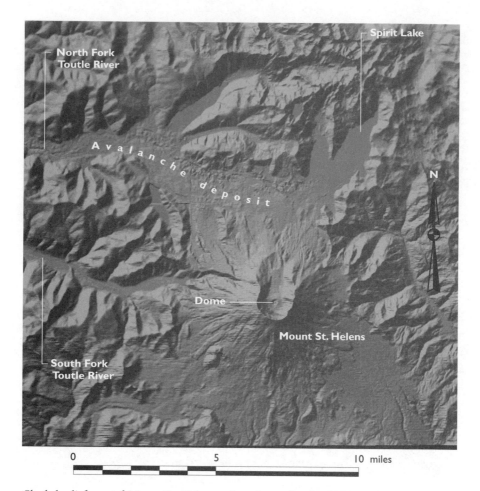

Shaded-relief map of Mount St. Helens and environs after the 1980 eruption.

St. Helens for two reasons. First, it was known to have erupted frequently in fairly recent times (the most recent significant eruption was
in 1857). Second, the volcano is close to populated areas; the village of
Cougar is about 12 miles from the summit, and the Portland/
Vancouver area is only about 45 miles away, albeit on what is usually the
upwind side of the volcano.

Mount St. Helens is a very young volcano—less than one-tenth the
age of Mount Mazama. Since its birth about 40,000 years ago, it has had
nine periods of eruptive activity, separated by dormant periods lasting
from 200 to as much as 15,000 years. The mountain as we now know

it, however, was constructed largely over the past 2200 years, especially during the Castle Creek period (about 2200 to 1600 years ago). During this period, numerous eruptions of andesitic, basaltic, and dacitic magmas formed the mountain's classic stratovolcanic cone. Another much briefer eruptive period (the Kalama period) happened about 400 years ago. Frequent eruptions of magmas of varying compositions raised the summit to about the height it had been before the 1980 eruption.

Even before 1980, the geological record of Mount St. Helens had spoken volumes about the hazards posed by the volcano. During some periods in the mountain's past, there had been repeated explosive eruptions, which ejected from 0.025 to 0.75 cubic miles of material. One of the eruptions in the Kalama period blanketed northeastern Washington and the adjacent portions of Canada in ash. The eruptions were often accompanied by mudflows, avalanches, and pyroclastic flows. All the nearby river valleys had been exposed to large-scale mudflows and debris flows that extended as far as 50 miles downstream. Spirit Lake, the once-beautiful lake on the volcano's northern slopes, was formed about 4000 years ago when a huge mudflow cut off the headwaters of the North Fork of the Toutle River.

Two years before the 1980 eruption, Mullineaux and Crandell published an official USGS bulletin outlining the possible hazards associated with a future eruption of Mount St. Helens. Based on the volcano's previous behavior, the publication described the areas likely to be affected by lava, pyroclastic flows, mudflows, and ash falls. This "Blue Book" was a useful resource for emergency planners, but it also had its limitations—at least two events during the eruption turned out to be far larger than anything the mountain had produced in earlier periods. In addition, the distant ashfalls were much more widespread and disruptive than anticipated. Thus, the past provided only a partial view into the future, a view that led to an underestimate of the potential hazard.

The first event in the 1980 eruptive sequence was a moderate earthquake that rattled the area around the mountain on March 20. The earthquake attracted little attention. Portland's newspaper, the *Oregonian,* which later made Mount St. Helens its lead story for weeks on end, mentioned the earthquake in the briefest of reports on an inside page. But the following day, officials of the Gifford Pinchot National Forest (the area that includes Mount St. Helens) called the USGS Volcanic Hazards Project in Denver, where Mullineaux and Crandell were based, to inform them of the earthquake. By March 24, earthquakes were occurring at frequent intervals, and on March 25 Mullineaux flew to Vancouver, Washington, to head an Emergency Coordination Center.

The first eruption was on March 27. In the early afternoon, a detonation was heard, and a cloud of smoke and debris rose about 6500 feet above the volcano. The eruption opened a small crater just north of the summit of the mountain. In addition, a fracture appeared, running east–west across the very summit of the mountain. The fracture suggested that there was a weakness in the north side of the volcano's edifice.

The eruption, although extremely minor, succeeded in focusing public attention on the mountain. The geologists were besieged with requests for information; for predictions; and for advice about road closures, evacuations, and so forth. Their answers were quite varied, even contradictory. David Johnston, who was already at his 4500-foot-high observation post, told the *Oregonian* that the mountain could erupt into a "glowing avalanche" and that a flow of rocks, ash, and molten lava could travel down the slopes at speeds of up to 100 miles per hour. "This is an extremely dangerous place to be," he said. "If it were to erupt right now, we would die. We're standing next to a dynamite keg and the fuse is lit. We just don't know how long the fuse is." The senior geologists were much more conservative in their predictions. Crandell, informed of Johnston's comments, explicitly disagreed with them, telling the newspaper that an eruption, if it happened at all, would likely consist only of a blanket of ash or pumice.

During the three weeks after the initial eruption on March 27, further minor eruptions took place. A second crater formed, and at night pale blue flames were seen within both craters. By mid-April, the two craters had coalesced to form a single crater about 400 yards across. The transverse fracture was by now a dramatic feature of the mountaintop: Johnston's measurements, made by aiming a laser at the mountain, had shown that the north side had sagged slightly, turning the fracture into a cliff some tens of feet high. Furthermore, it was apparent that the north side of the mountain was bulging outward. By early May, the bulge had extended over 300 feet northward and was continuing to expand at a phenomenal rate—about 6 feet per day. Earthquakes were felt hourly. In hindsight, it is obvious that these preliminary events could not have continued for long before the mountain lost its balance.

During the precursory period, many precautions were taken to minimize the risks to the local population and to visitors. The area around the volcano was divided into various zones according to the estimated degree of hazard. Roadblocks were set up by the Forest Service, and people living in the most hazardous areas were urged to evacuate. Most of these people did so, although some refused. Notable among the latter was 84-year-old Harry Truman, who ran a lodge at Spirit Lake on the north side of the volcano. He became something of a media darling on account of his feisty disregard for the imminent danger. "They think the mountain will do ol' Truman in," he told the *Oregonian,* "and one day he'll come floatin' down the river all bloated. That's a lot of ———. I don't know nothin' about her innards, but I know the outside of that mountain better than any man. That mountain won't hurt ol' Truman. I'll be around for another 20 years." He would be proved wrong.

To minimize the risks from mudflows, engineers lowered the water level in the Swift Dam reservoir, located on the headwaters of the Lewis River, just a few miles south of the summit. The main drainages on the northwest side of the mountain, however, the North and South Forks of the Toutle River, were not dammed. Even if they had been, no dam could have withstood the enormous mudflows that did eventually rush down these valleys.

On April 1, scientists monitoring seismometers that had been installed near the volcano noticed that Mount St. Helens had begun to "hum," if this is an appropriate word to describe the very low-pitched, steady oscillations that the seismometers were recording. Known technically as harmonic tremor, these deep-toned signals are recorded only at active volcanoes, and they are believed to be generated by magma rising within the volcano's inner conduits. The tremors continued on and off for 12 days. These ominous signals, along with other phenomena such as the ongoing small eruptions and the north-face bulge, led to a consensus among the scientists that there was a grave risk of a major eruption and/or a large landslide on the north face of the volcano.

In light of these warnings, Governor Dixy Lee Ray (herself a scientist by training) proclaimed a state of emergency on April 3. Among other things, she ordered National Guard personnel to staff the roadblocks. For a while, this had a certain effect. Nevertheless, there was a definite credibility problem, fueled in part by conflicting statements from scientists. On Wednesday, April 9, for example, the *Oregonian* reported that "some geologists were saying that the volcano had stabilized, even though one geologist said that Tuesday was its most active day so far." Soon, evading the roadblocks became a popular sport: people with four-wheel-drive vehicles could easily find a way around them, especially if they made use of the fire-trail maps that enterprising locals were selling by the roadside.

People who owned cabins on the mountain were particularly eager to get to them. A group of people who owned cabins in the Spirit Lake area were so persistent that Governor Ray eventually issued a special order allowing them to enter the Red Zone for a brief period. They did so on May 17, less than 24 hours before the main eruption. Other homeowners were scheduled to visit their property at 10 the following morning, but by that time there was no longer any property to visit.

Another exception was made for the Weyerhaeuser lumbering crews, who faced a prolonged loss of livelihood if the entire forest were closed to them. As it turned out, the climactic eruption came on a Sunday; if it had come on a weekday, hundreds of lumber workers would have been within the blast zone, and few if any would have survived.

In late April, the eruptions temporarily ceased, but they resumed on May 7. During none of the precursor eruptions, however, were there any signs of magma—all the ejected material was older rock from the mountaintop itself, kicked out by steam heated by the magma rising below. The volcano was clearing its throat, as it were.

The Eruption

At 8.32 A.M. on May 18, the bulging north face of Mount St. Helens finally reached the point of instability, and the result was a landslide. This was the trigger for everything that followed—a total release of energy roughly equivalent to the detonation of one Hiroshima-sized atomic bomb every second over the ensuing nine hours.

It so happened that two geologists, Keith and Dorothy Stoffel, were at that very moment watching the mountain from a light plane directly above the summit. Dorothy had a fear of flying, and this was her first flight in a small plane. She was just growing comfortable with the experience when all hell broke loose below the plane. The Stoffels later described what they saw:

As we were looking directly down on the summit crater, everything
north of a line drawn east–west across the northern side of the summit
crater began to move as one gigantic mass. The nature of the movement was eerie, like nothing we had ever seen before. The entire mass
began to ripple and churn up, without moving laterally. Then the entire
north side of the summit began sliding to the north along a deep-seat-
ed plane. We were amazed and excited with the realization that we
were watching this landslide of unbelievable proportions slide down the
north side of the mountain towards Spirit Lake. We took pictures of
this slide sequence occurring but before we could snap off more than a
few pictures, a huge explosion blasted out of the detachment plane. We
neither felt nor heard a thing, even though we were just east of the
summit at the time.

They did not have time to watch events develop further, because
the blast cloud rapidly threatened to envelop their plane. Although the
pilot, 23-year-old Bruce Judson, turned the plane and flew away at top
speed, the blast still gained on them. Judson was able to accelerate by
putting the aircraft into a steep dive; by heading southward through a
gap in the blast cloud, they just escaped—"by microseconds," as the scientists put it.

The avalanche was one of the largest in recorded history: about
three-quarters of a cubic mile of rock and ice rushed down the slope at
a speed of up to 150 miles per hour. One part of the avalanche overran
Spirit Lake, killing Harry Truman and raising an 800-foot-high wave
that rushed up the slope of the opposite shore. As the water washed
back into the lake basin, it carried with it tens of thousands of tree
trunks that completely clogged the lake's surface. Meanwhile, the bulk
of the avalanche turned left into the valley of the North Fork of the
Toutle River; 14 miles of the valley were filled with debris to a width
of a mile and an average depth of 150 feet.

At about the same moment as the avalanche began, an earthquake—the strongest of the entire year—was recorded by seismographs
that had been placed around the mountain. In the scientific post-
mortems that followed the eruption, it was assumed that this earthquake
was of volcanic origin and had triggered the slide. But later analysis by
Hiroo Kanamori of Caltech showed that the earthquake was actually
produced by the detachment of the mountain's north face; in other
words, it was the result, not the cause, of the landslide.

Rapid though the avalanche's progression was, it could not keep
pace with the ensuing blast. The detachment of the mountain's north

face had exposed the edge of the bubble of magma that lay concealed within the mountain. Suddenly released from its confining pressure, the steam within the frothy magma ripped the magma apart. The resulting explosion sent a blast cloud northward that overtook the descending landslide only a mile or two north of the summit.

It was the northward-directed blast that killed David Johnston and many others. A volunteer radio operator, Gerald Martin, who was stationed a couple of miles farther to the north, had time to send a slightly longer message than Johnston before he also died: "The whole north side is giving way. It's consuming the USGS people [Johnston's post] and it's gonna get me." This message was relayed to the duty officer at the command center of the Washington State Department of Emergency Services.

The total area of land devastated by the blast was about 250 square miles, all of it on the north side of the mountain. Near the mountain, nothing but bedrock remained. Farther away, every tree was knocked down and left as a denuded trunk, its bark, branches, and foliage stripped away. An hour later, climbers on Mount Adams, 35 miles away, were amazed by the sight of scorched greenery raining out of the sky.

A few people within the blast zone lived to recount their experiences. One man was felling timber with three companions, about 12 miles northwest of the summit. He heard a "horrible crashing, crunching, grinding sound" coming through the trees. It suddenly became totally dark, extremely hot, and difficult to breathe. He was knocked down, and when he stood up again, his back, mouth, and throat were seared by the heat. After about two minutes, he could begin to see again; he found that every tree had been knocked down and that everything was covered with about a foot of ash. The four men, although burned over nearly half their bodies, started to walk out of the forest through the ash and fallen trees. After 8 miles of walking, three of them were rescued by helicopter, but only one of the three survived his burns.

A few people managed to escape in vehicles. A couple who were in a car at a roadblock on the road to Spirit Lake, 11 miles northwest of the mountain, described the front of the blast cloud as a "big black inky waterfall" or like "boiling oil with huge bubbles" that approached them at high speed. They drove west along the North Fork Toutle Valley road at about 100 miles per hour; the cloud almost reached them at one point, but eventually they began to outrun it and reached safety.

Three people were overrun by the blast cloud as they were sitting in their pickup truck, 12 miles north of the mountain. The approaching cloud "looked like a boiling mass of rock—and just as high as you could see." They saw entire trees picked up and carried into the air. When the cloud reached them, it became totally dark and very hot—although paradoxically, chunks of ice also fell onto the vehicle. The side of the truck facing the mountain was extensively damaged, plastic trim was deformed by the heat, and a Styrofoam cooler in the truck bed melted. The people themselves received burns that had not fully healed three months later.

Even people outside the restricted zone were overtaken by the blast. Bruce and Sue Nelson were camping on Green River, more than 30 miles from the volcano. "There was no sound, no warning," Bruce Nelson told the *Oregonian*. "Suddenly, we found ourselves buried in debris from falling trees and ash. It took us 10 minutes to dig our way out. We didn't know which way was up. The sky was black as night because of this black cloud of ash overhead." A friend had a broken leg, and they helped him to a cabin. Then they walked for about 15 miles until, just before nightfall, they saw a helicopter and were able to flag it down.

Besides the avalanche and the blast cloud, the area around the mountain was hit by pyroclastic flows and mudflows. The pyroclastic flows did little damage, because they struck areas that had already been totally devastated. But the mudflows were another matter. They formed from melted ice in the avalanche and water from Spirit Lake, then were mixed with pulverized rock and soil to the consistency of wet cement. These flows rushed down the river valleys, especially the North Fork of the Toutle River, destroying 27 bridges, obliterating roads, and completely filling the watercourses for scores of miles, exactly as Mullineaux and Crandell had foreseen in their Blue Book. The mud even reached the Columbia River: a Norwegian cargo ship, the *Hoegh Mascot,* ran aground in 18 feet of water, where the charted depth was 45 feet. It took six days of intensive dredging before the river could be reopened to shipping.

Although the northward-directed blast caused most of the casualties, it was not the main event. As the landslide exposed the edge of the dome of magma within the mountaintop, a deeper section of the mountain, containing almost the entire magma dome, also slumped downward—not to the base of the volcano, but far enough downward that the conduit supplying the magma dome from below was now open

Eruption of Mount St. Helens: the Plinian column viewed from the south.
(Robert M. Krimmel and Austin Post, USGS)

to the sky. The instant the confining pressure was removed, a gigantic vertical eruption commenced, the beginning of a classic Plinian column. Within 10 minutes, the column had ascended to a height of more than 65,000 feet, terrorizing the passengers and crew of an airliner that happened to be flying overhead. Constantly resupplied by fresh magma rising from the depths, the vertical eruption continued unabated for nine hours. About 0.05 cubic mile of magma was vented during that period. As the eruption continued, it expanded into a far greater volume of pumice, ash, and dust, which was carried eastward by the wind.

Day became darkest night as the cloud passed over eastern Washington. The ashfall brought all normal life to a stop. In cities near Mount St. Helens, such as Yakima, the ash fell as relatively coarse, sandlike material, but farther away it became finer and more difficult to deal with. The small town of Ritzville, 60 miles southwest of Spokane, was blanketed by about 3 inches of a material like talcum powder that choked any car or other machinery that people were unwise enough to start. The ash refused to be swept up—it just drifted around with every slight breeze. All roads were closed, and the town remained cut off from the world for six days, during which time the 1800 townspeople had to accommodate 2000 stranded travelers. The hay harvest in the entire area of the ashfall was ruined. The subsequent wheat crop, however, was so stimulated by nutrients in the ash that it produced a record yield (although the harvesting machinery was damaged by the abrasive ash particles).

The Washington State Department of Emergency Services (DES) was notified of the eruption by Gerald Martin's final message, as well as by messages from the Emergency Command Center in Vancouver during the next few minutes, but it took the agency nearly two hours to put out a statewide warning about the eruption. The reasons for this delay, and its consequences, were later analyzed in a study by Thomas Saarinen and James Sell. According to this study, the DES was a small, underfunded, ill-equipped agency of 25 employees run by a political appointee with no hazard management experience. When the warning messages came in, the on-site staff insisted on waiting for "official" confirmation from the USGS and the Forest Service, but the phone lines soon became jammed and no such confirmation came through. Therefore, they did not act on the radioed warnings, nor even on reports from the Seattle Weather Service that described flash floods and the trajectory of the eruption cloud. The eruption was on radio and television news an hour before the DES did anything; in fact, according to one DES official, the DES used the news reports as its major source of information. And when the DES eventually did try to activate the National Warning System, which was supposed to alert all the local emergency services automatically, the system did not work. The DES had to resort to sending teletype messages to the law enforcement agencies around the state. The National Warning System also failed to work after two subsequent eruptions (May 25 and June 12).

Some of the consequences of the delay were merely ludicrous: the police chief of one town that was being pelted with chunks of pumice called the county Department of Emergency Services, only to be firmly reassured that no eruption was in progress. Other consequences could have been more serious. The ashfall reached Yakima, where the National Guard's rescue helicopters were stationed, before the DES warning. If the Guard had waited for the official DES notification, the aircraft would have been grounded there. As it was, the Guard acted on its own and was able to get 19 of the 25 helicopters out of Yakima before the ashfall began. These helicopters eventually undertook several hundred sorties and rescued 124 people.

The failure to alert the downwind communities to the approaching eruption cloud did not cost lives, but it caused enormous psychological as well as economic problems. Although the Blue Book spelled out the ashfall hazards and gave some suggestions for dealing with them, the USGS, the Forest Service, and the DES had all failed to distribute the book to agencies in eastern Washington. Nor had they made any other significant effort to educate those communities. In fact, there had been an almost total focus on the hazards in the immediate neighborhood of the volcano. Thus deprived of both long-term and short-term warnings, and lacking any information about to how to deal with the unprecedented hazard, the citizens of eastern Washington were emotionally and intellectually unprepared as day turned to night and the air filled with choking ash. No one knew what health hazards might be posed by the ash. No one knew how to prevent motors from self-destructing or how to clear away the ash. No one had gathered their families together or laid in supplies. Communication was minimal, and each community had to improvise its own response to the emergency, often over a period of days.

The Aftermath

During the first day, no one had much of an idea what had happened to the mountain, because the ash clouds prevented a direct view of the scene. Indeed, all sorts of false stories were circulating. The *Oregonian,* for example, reported that Spirit Lake no longer existed. This report may have been based on the misinterpretation of data from airborne radar; the usual radar reflection from the water was not detected, but this was in fact due to the mass of logs that clogged the lake's surface.

When the air cleared on the following morning, pilots flying over the mountain were greeted by an extraordinary sight. The uppermost thousand feet of the mountain had been completely removed; in its place was a vast, rocky amphitheater, open to the north. Where there had been verdant forest and brilliant waters there was now nothing but scoured rock, millions of fallen tree trunks, and mud-choked river valleys. Everything was ashen gray. Scores of people were dead, buried who-knew-where among the debris. And the damage to the wildlife population was incalculable.

One thing that was nowhere to be seen was lava. The highly viscous dacitic magma—about 0.05 cubic mile of it—had simply blasted itself out of the mountain as soon as it reached the surface. But after two further explosive eruptions (on May 25 and June 12), a dome of vapor-poor lava did finally rise from the floor of the amphitheater. It soon hardened, plugging the volcano's vent. The plug was blown open by eruptions on June 22 and August 7, and a new lava dome appeared on August 8. Another eruption blasted through that dome on October 17, but a new dome oozed up the following day—and so it went on through the end of the year. All these lesser eruptions were successfully predicted, and scientists, lumber workers, and others were safely evacuated. Over the succeeding years, the lava dome has gradually grown to about 800 feet in height.

The total short-term losses to the Washington economy have been estimated at about $900 million. This amount pales into insignificance compared with some more recent natural disasters such as Hurricane Andrew and the Northridge earthquake. Of course, the major factor limiting the economic losses was that most of the affected land was undeveloped. The Portland metropolitan area, although close to the volcano, was spared any damage except during the June 12 eruption, when a northeast wind deposited a thin film of ash on the city.

The Future

Most of the principal volcanoes of the Cascade Range are capable of erupting with little notice. The most threatening, on the basis of their

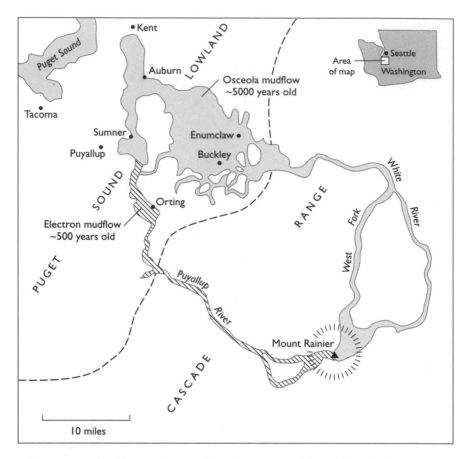

The areas inundated by two large mudflows that originated from Mount Rainier in the past. (Adapted from a drawing by C. D. Miller)

recent activity, are Lassen, Shasta, Hood, Adams, Rainier, Baker, and of course St. Helens.

There are several reasons for particular concern about Mount Rainier, the glacier-covered volcano that, at 14,410 feet, towers above the other peaks of the Cascade Range. Rainier's past eruptions have been of medium-silica andesite and have not been sufficiently explosive to sustain Plinian columns. Nevertheless, some of Rainier's eruptions have been dramatic in their consequences. The volcano lost about 2000 feet of altitude during a single eruption that took place about 5700 years ago. The avalanches resulting from this eruption gave rise to the Osceola mudflow, a wall of pulverized rock, debris, and glacial meltwater that

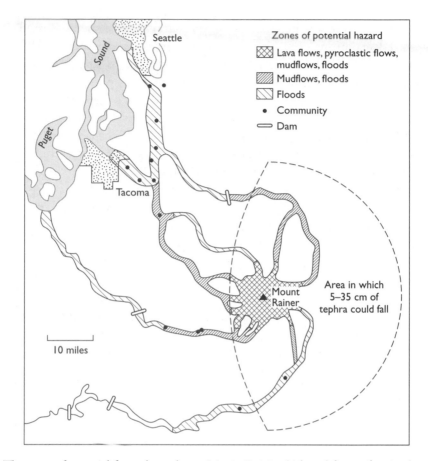

The zones of potential future hazard near Mount Rainier. (Adapted from a drawing by
C. D. Miller)

swept westward, inundating about 100 square miles of the Puget lowlands and even flowing directly into Puget Sound. Mount Rainier last
erupted during the early nineteenth century, but there are plenty of signs
of restless hibernation even today. Hot rocks, smoking vents (*fumaroles*),
and occasional steam blasts greet climbers as they approach the summit.

More than 100,000 people now live directly on top of the remains
of the Osceola and other mudflows. A total of about 2.5 million people inhabit the Seattle-Tacoma metropolitan area, barely 20 miles from
the mountain. The area is the high-technology center of the northwestern United States, as well as a major aircraft construction center. All
this would potentially be threatened by an eruption.

If there were adequate warning, the hazards from small or moderate-sized eruptions could be mitigated. For example, small mudflows could be held back by the many hydroelectric dams in the region, as long as there were time to lower the reservoirs. People could be evacuated from the most threatened regions, and public education about ashfalls could limit the psychological trauma as well as the economic hardship. Although a significant eruption would still cause a severe disruption of activity in the area, casualties might well be prevented.

But the hazards posed by Mount Rainier are not limited to eruptions. Perhaps even more dangerous is the hazard posed by a collapse of part of the volcano's edifice, even without an eruption. It doesn't take a geologist to realize how unstable much of the volcano's summit is. Made precarious by dome-building eruptions over the past few centuries, built of sloping slabs of crumbly rock separated by weaker layers of debris and ash, undermined by glacial scouring, and topped by a frosting of more than a cubic mile of snow and ice, much of the mountaintop resembles a crazily tilted wedding cake that waits only for some drunken guest to back into it.

In fact, that guest may already be approaching. For just a few kilometers from the volcano runs the Seattle fault (to be discussed in Chapter 6), which is believed capable of unleashing an earthquake larger than the one that rocked Los Angeles in 1994. It is a serious possibility that a major earthquake on the Seattle fault could trigger a landslide of enormous proportions. The slide could originate in Sunset Amphitheater, the huge cavity formed under the summit by a previous rockfall. Or it could involve the headwall of Point Success on the south side of the mountain, where layer after layer of ancient lava flows cling to the mountainside like congealed rivers of old wax clinging to a candlestick. Either way, the avalanches of rock and ice could turn into mudflows that run far down the neighboring valleys to the lowlands.

Scientists from the USGS and from the University of Washington keep an anxious watch on Mount Rainier. Seismologists keep track of the incessant tiny earthquakes that originate deep within the mountain. Volcanologists follow the changing patterns of heat and gas emission from the summit. Surveyors watch for bulges or rock movement. And government agencies develop response strategies—building, one hopes, on the lessons learned at Mount St. Helens.

Volcanoes similar to those of Cascadia are found around much of the Pacific Rim—hence the name Ring of Fire. Japan's Mount Fuji, celebrated in literature and art, has to be the best known, but almost every country with a Pacific coast has its share of stratovolcanoes, with their attendant threat of disaster: Mexico's Popocatepetl, Colombia's Nevado del Ruiz, New Zealand's Ruapehu, Pinatubo on the Philippine island of Luzon, Bezymianny on Russia's Kamchatka Peninsula, Augustine in Alaska. The famous volcanoes of the Mediterranean Sea are also stratovolcanoes: Vesuvius, which buried the Roman cities of Pompeii and Herculaneum; Etna, in Sicily, where the Greek philosopher Empedocles is said to have thrown himself to his death; Santorini, in the Aegean, whose eruption about 1600 B.C.E. may have initiated the decline of the Minoan civilization on Crete. Yet another string of stratovolcanoes lies in an arc in the southeastern Caribbean, near where the small Caribbean Plate is overriding its larger Atlantic neighbor. Martinique's Mount Pelée, which killed the 3000 inhabitants of St. Pierre in 1902, and the currently active Soufrière Hills volcano on Montserrat are examples. Another group of stratovolcanoes is scattered along the Indonesian island chain, where the Australian Plate is being subducted under the small Southeast Asian Plate; examples are Tambora, Toba, and Krakatau, which killed 36,000 people in 1883.

Let us now turn our attention to the region south of Cascadia, where the North American Plate is interacting with its western neighbor, the Pacific Plate, in a manner quite different from what happens in the Northwest: the North American Plate is sliding sideways against the Pacific Plate rather than overriding it. The main line of contact between the plates—the San Andreas fault system—is a source of earthquakes that have wreaked repeated havoc on California's cities and will undoubtedly continue to do so in the future. Its human impact has been so great, and so much has been learned by the scientific study of this one fault, that we devote three chapters to it. In the next chapter, we describe the earthquake that brought the fault into the world's consciousness, and we chronicle the gradual development of scientific ideas that seek to explain and predict its behavior.

3

THE GREAT DIVIDE: DISCOVERING THE SAN ANDREAS FAULT

The San Andreas fault is a star among faults, a seismological celebrity. Early one April morning, nearly a century ago, it destroyed a fabled city and thereby leapt from timeless anonymity to timeless myth. It is the fault that serves as metaphor for all that is dislocated, unstable, or apocalyptic about California and its people.

To study the San Andreas fault with the tools of modern science is a little like carbon dating the Shroud of Turin—there's a hint of sacrilege, of willfully deflating people's fond imaginings and replacing them with something all too matter-of-fact. Perhaps that's why so few

Californians visit the fault, even though it lies in plain view—a dramatic gash in the landscape that runs more than half the length of the state, from near Cape Mendocino in northern California to the Salton Sea, east of San Diego. Maybe they prefer the myth.

But the real San Andreas has its own romance. One of us (Kerry Sieh) has spent more than 20 years with the infamous fracture—measuring it, digging into it, reconstructing its history, and attempting to forecast its future. In this chapter, we'll try to present the San Andreas fault as it is: a window into the secret life of the Earth beneath our feet and an imminent hazard to Californians, but nonetheless a hazard whose dangers can be abated through knowledge and action.

The San Francisco Earthquake of April 18, 1906

There were many brave men before Agamemnon, and there were many great earthquakes on the San Andreas fault before the San Francisco earthquake. But in each case, history has preserved only a dim record of them. The San Andreas is believed to have ruptured at least twice during the nineteenth century. On December 8, 1812, an earthquake destroyed the new church at the Mission San Juan Capistrano, near the coast, halfway between Los Angeles and San Diego. Thirty-four Indian converts died in the rubble. Recent evidence, discussed in the next chapter, suggests that this "San Juan Capistrano earthquake" was actually caused by rupture of the San Andreas fault in the mountains northeast of Los Angeles. And on January 9, 1857, a very large earthquake on the San Andreas fault north and east of Los Angeles caused damage over much of south-central California and was felt as far away as Sacramento, Las Vegas, and northern Mexico. Because of the sparse population at the time, however, only two lives were lost.

As a result of these earthquakes, and of earthquakes that occurred on other faults during the nineteenth century, early inhabitants of California were well aware that they lived on terra infirma. During the 1890s Andrew Lawson, a geology professor at the University of California, Berkeley, had recognized the trace of a geological fault in the valley occupied by San Andreas Lake, south of San Francisco. In the months just before the great earthquake, he and his students traced the fault through much of the state, from Humboldt County in northern California to the Coachella Valley, north of the Salton Sea, in the south.

But there was little understanding of its significance. It took the San Francisco earthquake to draw public attention to the San Andreas and to provoke scientists to think about the underlying mechanism of its destructive ruptures.

The infamous earthquake struck at 12 minutes after five in the morning. Many agricultural workers were already in the fields; some of them heard a rumbling sound, and then, by the gray predawn light, they saw waves riding along the ground as if it had suddenly been transformed into a windswept sea. Some trees were shaken so severely that their crowns touched the ground, and the trunks of others snapped off entirely. Animals panicked; riders were thrown from their horses and cattle fell to the ground. Prodigious quantities of milk were spilled from churns. Water tanks overturned, and the ground itself failed in many places, torn by landslides, fissures, and sand-blows. Railroad lines buckled and water pipes broke, sometimes in hundreds of places. The shaking lasted about a minute.

The earthquake was dramatic in the countryside, but in the cities it was catastrophic. Many of the inhabitants of San Francisco, especially the well-to-do, were still asleep at the time, but they were rudely awakened by the first shock. The great Italian tenor, Enrico Caruso, who had had a triumphant success in *Carmen* the previous evening, rushed down to the lobby of the Palace Hotel, where he threw a tantrum worthy of the star he was, weeping floods of tears and swearing never to revisit a city that permitted such disorderly events. With the aid of an enormous bribe, he and his baritone costar, Antonio Scotti, were able to secure transportation out of the city.

Most of the citizens were stuck where they were. They milled about in the streets, wearing nightdresses or whatever they had thrown over themselves in their hurry to get out of their homes. Some reentered their homes to rescue the injured or to recover possessions. Some of these attempts proved fatal, as the almost continuous aftershocks brought down structures that had withstood the initial shaking.

The heavy masonry walls of City Hall collapsed, leaving the cupola perched incongruously on an iron latticework. Many other masonry buildings were destroyed. Still, the new skyscrapers along Market Street rode out the earthquake with little damage, and most wood-frame buildings—the vast majority of all the buildings in the city—survived, even if their chimneys cracked or fell. Only in the landfill areas near the waterfront did wooden buildings collapse in appreciable numbers.

San Francisco City Hall after the 1906 earthquake.

But if wood construction gave protection against the earthquake's direct effects, it was all too susceptible to a related scourge—fire. Within half an hour of the earthquake, at least 52 separate fires had broken out around the city. Other fires started later as householders rashly lit their hearths under damaged or missing chimneys. Soon, two major conflagrations were under way, one north, one south of Market Street. And there was no water.

In this perilous moment, the Commanding General of the Department of California, Frederick Funston, seized power. He rounded up several thousand soldiers and had a proclamation printed (by hand, since there was no electricity) and distributed; it threatened summary execution for looting. Funston's men did in fact shoot or bayonet many looters, when they themselves were not looting. Eventually, Funston's authority was given the appearance of legitimacy by the three San Francisco newspapers. They put out a joint extra edition, which was printed across the Bay in Oakland; it stated (untruthfully) that President Theodore Roosevelt had proclaimed martial law in the city.

Funston's main effort was aimed at dynamiting buildings in the path of the fires, with the hope of creating firebreaks wide enough to stop the passage of the flames. This effort failed repeatedly on the first day and on the following day, too. It failed because the buildings selected for demolition were too close to the oncoming fire and because the dynamite ran out, leaving the soldiers nothing but gunpowder, which set fire to buildings more readily than it collapsed them.

Block after block was consumed. Mechanics Hall, a huge warehouselike structure near City Hall that had become an impromptu hospital, morgue, and repository for art objects, burned to the ground. So did the Palace Hotel, despite the three-quarters of a million gallons of water stored on the premises for just such an emergency. The skyscrapers on Market Street, the tenements of Chinatown, the wharves—everything that could burn, did so. Only portions of Telegraph Hill and Russian Hill were spared.

When it became clear that the downtown area would be a total loss, the broad carriageway of Van Ness Avenue became the final line of resistance. Surely the fire could be stopped there and the residential areas to the west saved. Backfires were started in the elegant mansions along the east side of the avenue, but flames leaped across to the west side and ignited buildings there, too. It seemed as if the entire Western Addition was doomed, but thanks to heroic efforts by the firefighters, as well as by engineers who had partially restored the water supply, the fire was halted a block farther west, at Franklin Street.

The fire had raged for three days. A total of 28,188 buildings were destroyed in an area of 4.1 square miles. According to the Army's body count, 498 persons died in the earthquake or the fire; about four times as many were unaccounted for and probably also died. Property losses were $450 million in 1906 dollars (about $7 billion in current dollars). Recovery was slow and painful. And Caruso never did sing in San Francisco again.

The San Francisco earthquake devastated many other cities besides San Francisco. The City Hall of Santa Rosa, 50 miles to the north, was damaged even more severely than San Francisco's City Hall. Its domed tower fell onto the main building, crushing it completely. The recently completed buildings of Stanford University, 20 miles to the south of San Francisco, took terrible punishment; many of them were reduced to rubble. Buildings were damaged as far north as Humboldt Bay, near the Oregon border, and as far south as King City in Monterey County. The earthquake was felt over an extent of 175,000 square miles, from central Oregon to Los Angeles and from the coast to Winnemucca, Nevada. Even sailors at sea felt the shocks: some captains believed they had run aground, even though they were in deep water. The total number of deaths will never be known with certainty, but several thousand people must have perished.

Searching for the Cause

The 1906 earthquake was disastrous for San Francisco, but it was an enormous stimulus to science. The renowned geologist G. K. Gilbert expressed some of the scientific community's excitement in the first paragraph of an article he wrote about the earthquake, a year after its occurrence:

> It is the natural and legitimate ambition of a properly constituted geologist to see a glacier, witness an eruption and feel an earthquake. The glacier is always ready, awaiting his visit; the eruption has a course to run, and alacrity only is needed to catch its more important phases; but the earthquake, unheralded and brief, may elude him through his entire lifetime. It had been my fortune to experience only a single weak tremor, and I had, moreover, been tantalized by narrowly missing the great Inyo earthquake of 1872 and the Alaska earthquake of 1899. When, therefore, I was awakened in Berkeley on the eighteenth of April last by a tumult of motions and noises, it was with unalloyed pleasure that I became aware that a vigorous earthquake was in progress.

It soon became apparent to Gilbert and his fellow geologists that the earthquake had been caused by slippage of the Earth along an approximately 250-mile-long segment of the San Andreas fault, extending from Point Delgada in Humboldt County to San Juan Bautista, 80 miles southeast of San Francisco in San Benito County. Fences, pipelines, roads, and piers that crossed this segment of the fault were broken and offset horizontally, mostly by 10 to 15 feet. In all cases, the western side of the fault was displaced northward with respect to the eastern side. In general, there was little or no vertical offset.

If this kind of displacement were typical of all earthquakes on the San Andreas fault—and that later turned out to be the case—then one could describe the San Andreas fault as a *right-lateral strike-slip fault*. "Strike-slip"

Map of the San Andreas fault. The portions of the fault shown by solid lines produced the great earthquakes of 1857 and 1906. The arrows show the relative motion between the Pacific and North American plates.

because the slip is along the strike or horizontal dimension of the fault, and "right-lateral" because the direction of slip is such that an observer on one side of the fault sees the opposite side move to the right.

How could such a great movement of the Earth happen? Was the right-lateral slip a permanent characteristic of the San Andreas fault, or did it break in a different fashion on different occasions? What could be deduced from the 1906 earthquake about the likelihood of another earthquake at the same place or elsewhere in the state? Was there, in short, a pattern?

Strike-slip faults

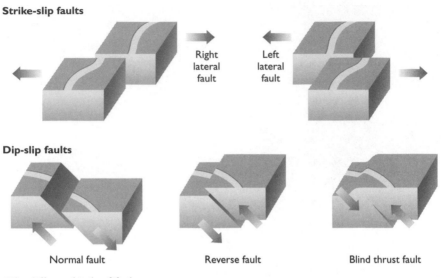

Dip-slip faults

Normal fault   Reverse fault   Blind thrust fault

The different kinds of faults.

A commission was set up to study the earthquake, and minute details of the event were recorded in two thick volumes that were published in 1908 and 1910. From the scientific point of view, one of the most interesting parts of the report was an analysis by an engineer from Johns Hopkins University, Harry Fielding Reid. Reid closely compared the results of land surveys that had been carried out in the San Francisco area during the nineteenth century with the results from surveys made after the 1906 earthquake. He found evidence that landmarks far to the west of the San Andreas fault (such as the lighthouse on the Farallon Islands, 25 miles offshore from the Golden Gate) were creeping gradually northward with respect to landmarks far to the east, such as the summit of Mount Diablo in Contra Costa County, 35 miles inland from San Francisco. Between surveys conducted around 1860 and surveys conducted about 20 years later, landmarks on the islands moved north by about four and a half feet with respect to Mount Diablo. By 1907, the islands had moved an additional six feet northward. Reid surmised that this gradual relative movement between land on the two sides of the fault distorted the terrain in the neighborhood of the fault, storing up energy in the form of elastic deformation or strain. When the strain exceeded the capacity of the rock to withstand it, the rock broke, allowing the terrain near the fracture to jump to a new, less strained position. Reid's "elastic rebound" theory has been the basis of all subsequent thinking about earthquakes.

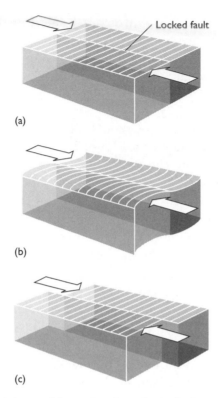

Reid's elastic-rebound theory of the earthquake cycle. At the beginning of the cycle (a), the fault is unstressed. Gradually, stresses deform the land on either side of the fault (b). Finally, the frictional resistance of the fault is overcome, and it slips to a new, unstressed position, causing an earthquake (c). (Adapted from Frank Press and Raymond Siever, Understanding Earth *2d. ed. New York: W. H. Freeman and Company)*

Reid's theory had been foreshadowed in earlier writings by other scientists. Gilbert himself put forward comparable ideas in the 1870s based on his observations of faults in eastern California and Utah (see Chapter 8). And Bunjiro Koto, of the University of Tokyo, came to similar conclusions as a result of his study of the catastrophic Mino-Owari earthquake of 1891, which killed more than 7000 people in central Honshu. Reid, however, was the first to recognize that these ideas could be the basis of a formal, mathematical description of the earthquake cycle.

Reid could not have understood the underlying cause of the slow deformation of the crust, for he had no notion of plate tectonics. But he did realize that, if the deformation went on in a regular, continuous way, similar ruptures might occur regularly and it might be possible to predict them. He estimated that the deformation that caused the San Francisco earthquake had accumulated over about a century and that similar earthquakes might therefore occur about this often. It was a good try: according to our current understanding, he was off by only a factor of two or three.

On the basis of this line of thought, Reid dreamed up an earthquake warning device. He proposed installing a row of piers, a kilometer apart, at right angles to the fault. Assuming that the rocks were now in their relaxed, unstrained position because of the recent rupture, the row of piers would gradually bend as the decades went by, revealing the distortion of the land as strain reaccumulated. When the curvature of the row reached the maximum elastic deformation that rock can undergo, a new earthquake would be imminent.

The piers were never constructed, and in fact they would not have worked as well as Reid had envisioned. For one thing, we now know that a fault rupture does not necessarily completely alleviate the accumulated strain, so the straight line of piers, even if constructed immediately after an earthquake, would not necessarily represent the relaxed state of the rock. Still, measurement of slow deformation of the ground has since become an important element in the study of earthquakes. Reid's analysis represented a major insight into earthquake dynamics, and it was the first quantitative attempt to apply this insight to the prediction of future shocks.

Running the Clock Backward

There was no instant consensus after the San Francisco earthquake that strike-slip ruptures were the San Andreas fault's characteristic mode of behavior. On the contrary, through the first half of the twentieth century, the general view among geologists was that, over the fault's lifetime, most of the displacements had been up or down the slope, or dip, of the fault, not sideways. In fact, dip-slip motion was thought to be the characteristic mode of rupture for *all* faults, not just the San Andreas. Strike-slip motion, if it occurred, was some kind of temporary adjustment or by-product.

The reasons for this deeply ingrained point of view were twofold. First, dip-slip faults clearly did exist; as Gilbert had shown 30 years earlier, they were responsible for the raising of many mountain ranges in the American West. Second, geologists had an explanation for dip-slip motion, which was the cooling and consequent shrinking of the Earth's interior over geological time. This shrinking, it was believed, tended to throw the Earth's brittle crust into folds and ridges, like the wrinkled skin of a dried-out apple. All this vertical displacement had to be accommodated by folding and by motion on dip-slip faults. Strike-slip motion, on the other hand, seemed to threaten a rearrangement of the Earth's geography, a rearrangement that didn't fit into anyone's world-view. So the 15-foot horizontal offset that accompanied the 1906 rupture was seen as a curiosity, not as part of a pattern.

In 1909 two geologists, Ralph Arnold and Harry Johnson, gave a lecture at the Geological Society of Washington describing their visit to a stretch of the San Andreas fault much farther south than the segment that ruptured in 1906. The place they visited was the Carrizo Plain, a hot, dry, uninviting piece of real estate tucked in between the Coast Ranges and the aptly named Temblor Range, about 120 miles north-west of Los Angeles (*temblor* is the Spanish word for a trembling or an earthquake shock). The San Andreas fault runs along the southwestern foot of the Temblor Range, and many dry streambeds, cut by the runoff from occasional winter storms, cross the fault on their way westward to a dry lake bed on the valley floor (see Color Plate 9). Arnold and Johnson showed lantern slides that depicted one of these streambeds: it descended to the fault, turned right, and followed the fault northwest-ward for 400 feet, then turned left and continued down the slope into the valley. What could be the reason for this 400-foot dogleg, they asked, if it were not a 400-foot right-lateral slippage of the fault that had offset the path of an originally straight streambed? Arnold and Johnson had no idea when this offset was produced, nor whether it was the result of one gigantic rupture or a chain of smaller ones. But it sug-gested that the strike-slip motion of 1906 was part of a larger pattern, not just a fluke.

Arnold and Johnson's lecture was probably greeted with skepti-cism. Certainly, it fell on deaf ears, and their observation was widely ignored. But today, with the advantage of aerial photography, the off-sets of this and many other streambeds in the Carrizo Plain are beyond debate. All the offsets are right-lateral, but the size of the offsets varies, as if the streambeds were initially cut at different times in the past and thus have been subjected to varying numbers of earthquakes. As we will see later, these streambeds hold important clues to the recent behavior of the fault.

For nearly half a century after the San Francisco earthquake, the consensus among geologists remained that strike-slip motion on the San Andreas fault was anomalous; most or all of its prehistoric motions, they agreed, had been dip-slip. In holding to this view, they had to circumnavigate an awkward and increasingly apparent fact about the fault: for much of its course, the rocks on either side of the fault are different. To take one example, 30 miles north of San Francisco the fault runs along the bottom of Tomales Bay, the narrow, rail-straight inlet that separates Point Reyes from the mainland. Point Reyes is granite, overlain in parts with shale (a marine sedimentary rock). The mainland here is Franciscan Formation, which is to say a ragbag of rock types, much of it derived from ancient oceanic crust. To explain this unlikely conjunction, Andrew Lawson (the Berkeley professor who had mapped the San Andreas fault before the 1906 earthquake) proposed the following history. Originally, both sides of the fault consisted of Franciscan rock overlying a granite basement. Then the west side of the fault was elevated many thousands of feet by dip-slip motion. The Franciscan rock on that side was completely eroded away. Next, the west side sank again, to below sea level, and shale was deposited on top of the granite. Finally, the west side rose up once more, leading to the current topography.

As if this kind of explanation didn't involve enough complexities, more had to be added. For example, there are a number of sites where the geological differences between the two sides of the fault reverse themselves within a few miles: at one site, it's X on the west and Y on the east, and then a short distance away, it's Y on the west and X on the east. To explain this observation, geologists had to invoke a scissors pattern of motion: uplift and subsidence at one site, subsidence and uplift at the other.

A new era of thought began, very tentatively, in 1926, with a proposal by a Yale-trained geologist named Levi Noble. As a wedding present, Noble received a most appropriate gift—the San Andreas fault! A mile or two of it, at least, wrapped up in a charming ranch in the northern foothills of the San Gabriel Mountains. So what could be more natural than for Noble to examine the geology of the fault in this region? Near his home, at a place on the southwestern side of the fault called the Devil's Punchbowl, Noble found and mapped a picturesque rock formation. These contorted rocks consist of sands and gravels deposited by an ancestral river. To Noble's surprise, he found a very similar rock formation on the other side of the fault, but about 20 miles to the southeast. This rock formation lies in the Cajon Amphitheater, where Interstate 15 now squeezes between the San Gabriel and San Bernardino ranges on its way from the Mojave Desert down to the Los Angeles basin. (Known as the Mormon Rocks, they have since served as a scenic backdrop for innumerable westerns.) "It thus appears possible," Noble wrote in a report to his sponsoring organization, the Carnegie Institution of Washington, "that horizontal movements along the fault have dragged the rock masses north of the fault to the southeast."

Noble's suggestion of a 24-mile offset, put forward very tentatively, was not received with enthusiasm. And, in fact, he was wrong, although it took 50 years to prove him so. In the 1970s, Michael Woodburne, a paleontologist at the University of California, Riverside, compared the fossils in the Punchbowl Formation and the Mormon Rocks (see Color Plate 14). They were of different ages—not greatly different, but enough to prove that the two groups of rocks were never part of a single geological unit.

But, in Noble's time, his claim was disbelieved for the wrong reasons. Powerful voices spoke against the notion of extensive horizontal motion. Most prominent among these opponents was Nicholas Taliaferro, a geologist at the University of California. As late as 1938, Taliaferro declared that the total strike-slip motion on the San Andreas fault was "certainly less than a mile." To explain away some of the inconvenient data, Taliaferro claimed that the San Andreas fault was extremely young, and that, before it came into being, extensive motions had taken place on older faults in the same location.

Still, Noble's suggestion gradually opened the door to a whole new way of thinking about the fault. And during the 1930s and 1940s, as oil-prospecting geologists mapped more and more of the state in their search for black gold, the notion of horizontal offsets became ever more plausible. Two petroleum geologists, Mason Hill and Tom Dibblee, made many of the key observations and interpreted them. They not only described numerous apparent offsets along the San Andreas fault but also proposed that there was a systematic relationship between time and distance: the older the rocks that had been sundered by the fault, the greater the distance by which they were now displaced. Ironically, their work led them to criticize Noble for the exact opposite reason that Taliaferro had attacked him; they argued that the offset he had proposed, given the age of the rocks involved, was too small! Poor Noble—the offset he proposed was disbelieved by everyone and was finally proved wrong, yet his proposal was in the vanguard of modern thinking about the San Andreas fault.

Hill and Dibblee laid out many examples of offsets, most of them based on their own studies of the south-central portion of the fault. At the south end of the Carrizo Plain, for example, are layers of gravel—rock broken into small pebbles by the rivers of the Pleistocene epoch, perhaps half a million years ago. The pebbles are completely different on either side of the fault: on the west they're a hodgepodge of granite, gneiss, quartzite, limestone, black shale, and sandstone; on the east they're nearly all white shale. But slide the western hodgepodge 10 miles to the southeast, and it comes up against an assembly of all the parent rocks that those pebbles were made from. So something like half a million years ago, it seems, the fault was 10 miles "back" of where it is now.

As another example, about 6 million years ago, in the late Miocene epoch, the San Joaquin Valley (the southern part of California's Central Valley) was a shallow inland sea. Like a ring around a bathtub, the coastline of that period is visible today as a sharp transition from marine sediments to continental deposits; it runs around the southernmost foothills of the Sierra Nevada and meets the San Andreas fault not far from where Interstate 5 exits the valley. Here, where eighteen-wheelers now gear down for the torturous ascent of the "Grapevine," camels may once have romped in the surf. But on the western side of the fault near this spot, there's no coastline to be found—it's all continental rock. To find the continuation of the marine-continental transition, one has to go 65 miles to the north. So the side of the bathtub, ring and all, has slid 65 miles in 6 million years.

When Hill and Dibblee looked at even older rocks from the Eocene epoch (which ended about 37 million years ago), they found much greater displacements. Eocene rocks on the east side of the fault at the south end of the San Joaquin Valley seemed to match up with rocks on the west side of the fault in the mountains near Santa Cruz—a slippage of about 190 miles!

Where, then, does it all end—or rather begin? The answer had to await the advent of the theory of plate tectonics in the 1960s. It then became clear that the San Andreas fault is part of the boundary between the Pacific Plate and the North American Plate. To understand why strike-slip motion is taking place along this part of the plate boundary, we need to expand on the history of the Pacific Ocean, a topic already touched on in Chapter 1.

We mentioned in that chapter that the Pacific was originally created by the activity of a mid-ocean ridge that ran roughly north-south; the ridge spun off new ocean floor to the west and to the east. For millions of years, the western half of the new ocean floor was being recycled at a subduction zone near the eastern shores of Asia, while the eastern half, known as the Farallon Plate, was being subducted near the western shores of the Americas. Starting about 180 million years ago, however, this expansion became very asymmetrical, because by that time the Atlantic Ocean began to open up, pushing the Americas westward. Because of this westward motion of the continents, subduction along their western margins was speeded up, and the Farallon Plate was now being consumed faster than it was being created.

Just as a reasonably fit person can run up a "down" escalator faster than the escalator descends and thus eventually reach the escalator's origin at the upper landing, so North America eventually overrode the advancing Farallon Plate so far that it reached the plate's source, the mid-ocean ridge. Because of the curvature of the continental margin, this contact did not take place all at once. The first contact was made by part of the coastline of southern California at the approximate level of Santa Barbara, about 28–30 million years ago.

Once a portion of North America had overridden the ridge, that portion was now no longer up against the Farallon Plate but against the other, western half of the ocean floor—the part we now call the Pacific Plate. But this piece of ocean floor was moving in the other direction, westward—or, to be more exact, northwestward, toward Alaska and the Aleutian Islands. A strike-slip fault must have immediately developed, allowing the Pacific Plate to continue its northwestward motion relative to the North American continent. The fault took the place of the length of mid-ocean ridge that had been overridden. At first it was just a few miles long, but as more and more of the continent reached the ridge, the fault became longer and longer.

That original strike-slip fault (or set of faults) was not the San Andreas fault. We know that because the San Andreas fault in south-central California is far inland, not at the continental margin. Also, the San Andreas fault has accumulated no more than about 190 miles of strike-slip motion over its entire lifetime, whereas the total sliding

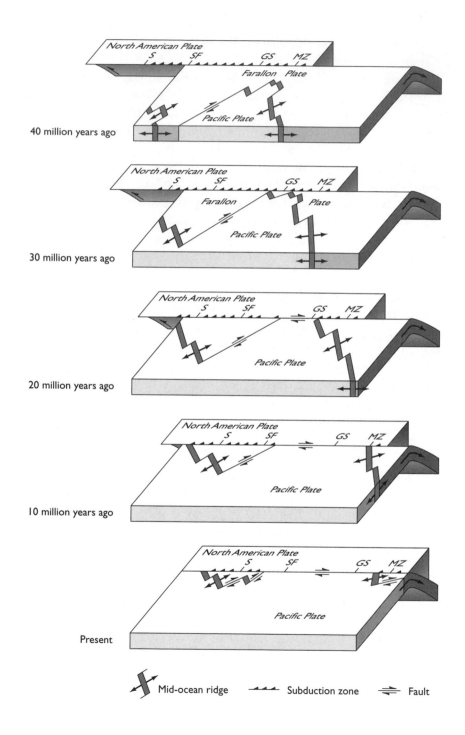

The origin and expansion of the San Andreas fault system as a transform plate boundary between the Pacific and North American plates. This is a simplification, in that the location of the transform fault has probably shifted at least once during its evolution. S = Seattle; SF = San Francisco; GS = Guaymas, Mexico; MZ = Mazatlan, Mexico. (Adapted from W. Irwin, USGS Professional Paper 1515)

motion between the Pacific and North American plates since the two plates made contact (based on the analysis of seafloor magnetism) has been 700–900 miles. Strike-slip motion on other parallel fault systems, and possibly rotations of large blocks of crust, must account for the approximately 500- to 600-mile difference.

Geologists by no means agree about the details of this early history. Most likely, however, the first faults were near or off the coast of southern California. For perhaps 10–12 million years, the plate boundary, although gradually lengthening, remained at or near the margin of the continent. Around 18 million years ago, however, the bulk of the relative motion was transferred inland, forming a "proto–San Andreas fault." Thus, the land between the fault and the coast was donated, as it were, by the North American Plate to the Pacific Plate. (Since most of this land was originally formed from chunks gouged out of the Farallon Plate during its subduction, one could consider this a return of property rather than an original gift.)

The proto–San Andreas fault probably ran from the coast, in the general region of Monterey Bay, along the course of the modern fault as far as the south end of the San Joaquin Valley. Farther south, however, it seems to have taken a different route, first to the west and then to the east of the modern fault, ending up somewhere east of the present site of the Salton Sea.

About 5 million years ago, the long sliver of coast that we now know as Baja California was split off from the mainland of Mexico by an extension of the East Pacific Rise and started drifting northwestward with the Pacific Plate. The spreading center that now runs up the Gulf of California to the Salton Sea is broken into many short segments by long transform faults. Collectively, the transform faults permit Baja California to move northwestward with respect to the Mexican mainland. These transform faults can be considered an extension of the San Andreas fault system, which thus runs all the way from Cape Mendocino to the level of Mazatlan, Mexico. The spreading center's northern limit is the deep trough now occupied by California's Imperial Valley and the Salton Sea.

During the same period, the continuing westward drift of the North American Plate brought more and more of the continent up to the mid-ocean ridge, causing the San Andreas fault to extend northward. This younger portion of the fault stayed fairly close to the coastline all the way to its present terminus at Cape Mendocino.

If this process continues in the future, the San Andreas fault will become even longer, and the remnants of the Farallon Plate to the north (the Juan de Fuca and Gorda plates) and to the south (the Cocos Plate) will shrink further and eventually be swallowed up. We cannot tell how long the San Andreas fault will continue to be active. The plates will eventually change their relative motions, and even before then the line of slippage may shift to another fault system. If the current displacement rate continues, however, San Francisco will become an outlying suburb of Los Angeles around the year 16,000,000. That's if urban sprawl doesn't accomplish the same thing in the meantime.

4

IN THE TRENCHES:
UNEARTHING THE
SEISMIC HISTORY OF
THE SAN ANDREAS FAULT

From what we know of the long-term history of the San Andreas fault, we can be confident that there will be many more right-lateral strike-slip ruptures of the fault and that many of these ruptures will cause damaging earthquakes. But that is all we can be sure of. To go further—to say, for example, "An earthquake of about magnitude 7.0 will probably occur near Hollister about 50 years from now"—would require knowledge of the fault at quite another level of detail. We would need, at the very least, a blow-by-blow description of the fault's recent behavior. What was the date of the most recent rupture of each segment of the fault, how long was the sector that broke, how far did it slip, and how strong was the shaking it caused? And what about the preceding rupture, and the one before that? By establishing a detailed chronology of slippages for every region along the fault, we might be able to discern an underlying pattern—the basic machinery of the earthquake cycle—that would tell us what to expect in the future. If, for example, such a chronology showed that Reid's hypothesis was correct (regular, complete release of accumulated strain about every hundred years), we could predict a repeat of the 1906 San Francisco earthquake around the year 2006. And a chronology of slippages would tell us whether there is a spatial pattern: might it be, for example, that ruptures follow each other in an orderly sequence along the length of the fault, setting each other off like a row of dominoes? If so, knowing the location of the last rupture would help us to predict the location of the next one.

The 1857 Earthquake

It is this kind of questioning that has motivated Kerry Sieh to study the San Andreas fault over the past 20 years. In the mid-1970s, when he was a graduate student in geology at Stanford University, Sieh decided to focus on the last great earthquake on the San Andreas fault prior to the 1906 San Francisco earthquake. This was the Fort Tejon earthquake of 1857, which resulted from the rupture of a long stretch of the fault in south-central California. His idea was to research the earthquake in two principal ways. First, he would collect all the scattered historical references to the earthquake, sketchy though they might be, to define as accurately as possible the region that had been involved in the earthquake, as well as the strength of the shaking. Second, he would examine the fault itself to find out precisely what part of it had ruptured and how great the ground displacement had been. In addition, he hoped that he might be able to establish the overall rate at which the San Andreas fault was slipping by dating very old offsets with the radiocarbon method. If so, he might be able to calculate how often 1857-type earthquakes occurred and, therefore, when the next one could be expected.

As soon as he began the project, chance observations, new ideas, and discussions with other scientists led him into several new avenues of research, and he ended up working on the San Andreas fault for two decades, rather than the two years that he had planned on at the outset.

Sieh's thesis proposal wasn't too popular with his professors at Stanford. For one thing, there was little precedent for a study of this kind. Brian Atwater's work on the Cascadia subduction zone, for example, was still a decade in the future. It was known that offsets caused by very recent ruptures could be detected and measured, as was done after the San Francisco earthquake. But it was not generally believed that ancient ruptures would leave a readable record. And even if there were evidence of such ruptures, it seemed unlikely that they could be precisely dated—least of all in the Carrizo Plain, where, aside from tumbleweed rolling by on the wind, it seemed that organic material suitable for radiocarbon analysis would be hard to come by.

There was another, unspoken objection to Sieh's proposal, which was that it was unworthy, even boring. For geology has always focused on the grand vistas: the raising of mountain ranges, the leveling of plains, and (since the 1960s) the movements of oceans and continents. Certainly, to see these vistas one had to lay a hammer to a particular piece of rock somewhere. But why study the last century or two of slip on the San Andreas fault, a structure whose far more ancient history had already been worked out? There was a literal superficiality to the proposal: Sieh was going to devote himself to the thin layer of compacted dirt that, in most geologists' eyes, serves only to obscure the ancient history recorded in the "real" rock below.

Still, Sieh got some encouragement from a more senior geologist at the USGS in nearby Menlo Park, Robert Wallace. During his own graduate student days in the late 1930s, Wallace had been a player in the long-running debate over the nature of the San Andreas fault. He had argued for miles of strike-slip motion, basing his argument in part on stream offsets. In 1968, in preparation for a conference about the fault, Wallace had measured more than 130 offsets in the Carrizo Plain. Ten of these offsets were about 30 feet in size, while the remainder were larger. Wallace surmised that the 30-foot offsets had been generated during the great earthquake of 1857, because that was the most recent large earthquake in this part of the state. The larger offsets presumably resulted from the combined offset of the 1857 event and one or more previous events. Thus, there was some inkling of a precedent for the kind of project that Sieh planned.

Sieh spent most of the summer of 1975 studying the fault. He covered about 200 miles of it, mostly on foot, sometimes by bicycle. He was looking especially for the smallest offsets of stream channels and other features, for such offsets were probably caused by the 1857 rupture. As it turned out, these offsets were greatest (about 30 feet) in the Carrizo Plain and much less (about 10 to 15 feet) through much of the southern part of the region that ruptured (the so-called Mojave segment).

That same summer, Sieh (and, independently, a Caltech undergraduate named Duncan Agnew) began to hunt down all the documentary records they could find that referred to the 1857 earthquake. Together, the field survey and the historical records showed that the rupture extended for at least 220 miles, from about the San Benito–Monterey County line south to Wrightwood, a resort town in the San Gabriel Mountains northwest of San Bernardino. And it was also possible to estimate the magnitude of the 1857 earthquake: it was a 7.8 or 7.9, about twice the size of the 1906 San Francisco earthquake, which was a 7.7.

Interestingly, the damage in the pueblo of Los Angeles was not especially severe: walls cracked, but no buildings collapsed. Thus it seems that a repetition of the 1857 rupture, even though it would extend to within 35 miles of the Los Angeles metropolitan area, would not cause extensive damage in the city, at least to low-rise buildings like the adobes that existed there in the mid-nineteenth century.

Wallace Creek

Sieh now had the 1857 earthquake in his bag, and with this study he had doubled the number of San Andreas earthquakes whose vital statistics were known. He could turn his attention to evidence for even earlier ruptures along the same south-central segment of the fault. He wanted to know how similar they were to the 1857 event and how often they had occurred.

In the fall of 1975 he began studying a site in the Carrizo Plain that he later came to name Wallace Creek, in honor of the man who had encouraged his research project. It was a streambed with a simple 420-foot dogleg, which seemed like a good candidate to work out the chronology. It was in fact the same offset stream that Arnold and

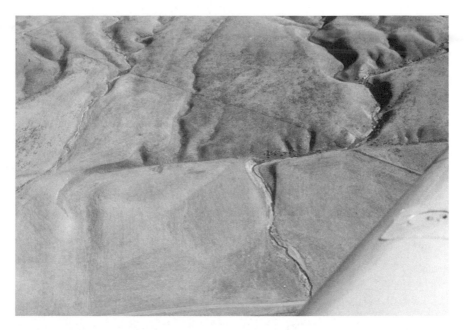

Looking northeastward across the San Andreas fault at Wallace Creek. (Kerry Sieh and Simon LeVay)

Johnson had described 66 years earlier, but at the time Sieh knew nothing of the earlier study.

Sieh found an old, abandoned channel that continued northward along the fault for 820 feet, and then turned left across the fault and headed down into the valley. Evidently this had been an earlier channel of Wallace Creek, but after it became displaced too far to the north, the creek cut itself a new route directly across the fault. Sieh realized that if he could establish the date when the creek cut its new channel, he could figure out the rate at which the fault had been slipping since then: it would be the offset of the present channel (420 feet) divided by the number of years that had elapsed.

Through sporadic efforts over several years, Sieh was able to fix the date at which the creek cut its new channel. He did this by looking for organic material, suitable for radiocarbon dating, that had been deposited in the bed of the abandoned channel. Before this channel was abandoned, it got longer with each successive rupture of the fault. New streambed was repeatedly added at the upstream end (marked X in the diagram). Therefore, the upstream end was the youngest portion of the

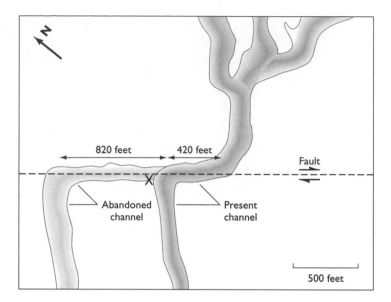

Diagram of Wallace Creek showing the current and abandoned channels and the site where samples for carbon dating were found (marked **X***). (Based on a drawing by R. Wallace)*

abandoned channel, and so any organic material deposited by the stream there could only slightly predate the cutting of the new channel. Equally, such material could not postdate the cutting of the new channel, because after that event no more water ran in the abandoned channel. Sieh did in fact find organic material at the point marked **X**, and it was radiocarbon dated at about 3800 years old. If the stream cut its new, direct path across the fault about 3800 years ago (or slightly more recently) and was now offset by 420 feet, the fault must have been slipping at an average rate of about 34 millimeters (1.3 inches) per year.

According to the most recent estimates, based on GPS measurements, the Pacific and North American plates are currently sliding past each other at a rate of about 49 millimeters (1.9 inches) per year. Thus the Carrizo segment of the San Andreas fault accommodates about 70 percent of the total relative motion between the Pacific and North American plates. The remaining 30 percent must be accommodated by slip on other faults on the coastal or inland sides of the San Andreas fault.

How many ruptures of the fault did it take to produce the 420-foot offset? If every rupture produced an offset of 30 feet, as the 1857 rup-

ture did, then it would take a total of about 13 ruptures. And 13 ruptures in 3800 years would mean an average interval between ruptures
of about 290 years. Thus, if the ruptures were regularly spaced in time,
the next great earthquake on the Carrizo segment would not be due
until 1857 plus 290 years, or around the year 2150.

For several years, Sieh believed that this was a more or less accurate
picture of the behavior of the fault. But direct evidence was lacking:
there were no reliable indicators of the actual dates or sizes of individual ruptures before 1857, so the crucial assumption, that earthquakes
were regularly spaced in time, was just that—an assumption. The
assumption began to fall apart as a result of research that Sieh conducted farther south along the fault.

Pallett Creek

In the fall of 1975, Sieh was already turning his attention to the Mojave
segment of the San Andreas fault, the stretch that runs across the high
desert near Palmdale and then up the eastern slopes of the San Gabriel
Mountains to Wrightwood. The reason for his interest in this segment
was as follows. He knew from his general reconstruction of the 1857
earthquake that the slip in that event had been much less along the
Mojave segment (about 10 to 15 feet) than in the Carrizo Plain (30
feet). He thought that the difference might allow him to distinguish
between two theoretical models of the earthquake cycle.

In the *characteristic-earthquake model,* great earthquakes repeat themselves in all their important properties. So the 1906 San Francisco
earthquake would have been preceded by umpteen very similar earthquakes on the northern part of the fault, and the 1857 earthquake
would have been preceded by umpteen very similar earthquakes on the
south-central part of the fault. If that model were correct, the Carrizo
and Mojave segments would always have broken together in the same
earthquakes, as they did in 1857, and on each occasion the Carrizo segment would have slipped by a two- or threefold greater distance than
the Mojave segment. The characteristic-earthquake model is really an
extension of the ideas of Harry Reid, described in Chapter 3.

There's a bit of a problem with this model, however, as applied to
the southern San Andreas fault. If the long-term slip rate is greater in the
Carrizo segment than in the Mojave segment, something has to give at

the border between the segments. On the coastal side of the fault, where the land is moving northward, a yawning chasm would gradually open up between the two segments, as the Mojave segment got "left behind." On the inland side, where the land is moving southward, there would be a piling up of the terrain at the border as the Carrizo segment overtook its slower moving neighbor. In fact, though, there is no evidence of such disturbance of the terrain between the two segments.

In an alternative model, which we might call the slip-patch model, great earthquakes don't necessarily repeat themselves, but a given segment of a fault tends to slip by a constant distance on each occasion. This distance might depend on the characteristics of the rock or the geometry of the fault at that locality. For example, the Mojave segment might slip by 15 feet and the Carrizo segment by 30 feet in every earthquake, but the Mojave segment might break more often than the Carrizo segment. So sometimes the Mojave and Carrizo segments would break together in large, 1857-style earthquakes, and sometimes the Mojave segment would break by itself in smaller "catch-up" quakes. Thus the slip rate for the two segments, averaged over many years, could be the same.

To decide between these two models (which don't exhaust the range of possibilities, of course), Sieh needed to determine the average slip rate along the Mojave segment and see whether it was in fact the same as along the Carrizo segment. So he looked for a big stream offset in the Mojave segment that he might be able to date. He found what he was looking for—or so he thought—at Pallett Creek, 35 miles northeast of Los Angeles.

Pallett Creek is a real stream—the kind with water in it. It flows down from the small ski resorts of the San Gabriel Mountains, crosses the San Andreas fault, and wanders on towards Pearblossom and the high desert. The water sustains a few willows and cottonwood trees, making the site far more hospitable than the bleak landscape at Wallace Creek. It is the kind of place where one might be tempted to stop for a brook-side picnic on a hot summer afternoon. What aroused Sieh's interest, however, was that the stream made a 425-foot rightward dogleg as it crossed the fault. In the course of making this dogleg, the stream had cut a 30-foot-deep gorge through the layers of peat and sediment that covered the area.

At first, Sieh assumed that this dogleg was an offset caused by slip on the fault, just like the dogleg at Wallace Creek. And he expected to

Pallett Creek during major excavations in 1980. The light upper surface was a swamp until the beginning of the twentieth century, when the creek cut the gorge in which the trees are growing. (Kerry Sieh and Simon LeVay)

be able to date the dogleg with the radiocarbon method and thus establish the long-term slip rate for the Mojave segment and compare it with the Carrizo segment. So Sieh took a sample of peat from the uppermost layer of deposits in the wall of the gorge—the last layer to be laid down before the creek started cutting its gorge—and sent it for radiocarbon dating. To his surprise, it turned out to have an age of only 200 years. This meant that the gorge had only existed for that long. A 425-foot offset couldn't possibly have formed in a mere 200 years, so the dogleg hadn't been caused by slip on the fault at all. It was just a turn that the stream had made to go around a small hillock. It could as easily have been a leftward as a rightward dogleg.

Although disappointed by this result, Sieh realized that Pallett Creek represented a different kind of opportunity. Here was a 30-foot stack of sediments built up over a period of centuries or millennia. If he could find the fault trace as it cut through the layers, maybe he could identify disruptions of the layers caused by individual earthquakes during this period. Perhaps as many as dozens of earthquakes had left their signature in the soil. So he started excavating in the walls of the gorge, looking for signs of earthquakes past.

To understand how the history of ruptures is preserved in the stratigraphy (the pattern of layers in the soil), we can make an analogy with slicing a wedding cake—a simpler one than the volcanic cake we left teetering at the end of Chapter 2. Let's start by assuming that the cake has been made so skillfully that all the layers—jam, fruitcake, marzipan, and frosting—are perfectly flat and horizontal. If we sliced such a cake with a vertical knife cut (representing the earthquake fault) and then slid the two halves an inch or two horizontally against each other (representing the offset induced by a rupture of the fault), we would not disturb the stratigraphy at all: jam would still be up against jam, frosting against frosting, and so on. If we then made a second cut at right angles to the first (representing a trench cut across the fault, or the canyon cut by the creek) and looked at the layers of the cake, we'd see nothing unusual at the line where the original halves abutted each other—the fault. We might not even be able to see the junction line at all.

But real cakes have imperfections. Often, for example, the fruitcake portions rise in the middle during the baking, so the baker compensates by making the jam layers a little thicker around the perimeter of the cake. In this case, after we've slid the halves against each other, the layers won't line up perfectly. When we make our second cut and examine the line where the two original halves abut each other, we'll see a slight dislocation of the layers: in places, jam will be up against fruitcake, fruitcake against marzipan, and so forth.

So it is with fault ruptures. The ground is rarely completely flat, and so when the fault slips by many feet, there is usually some kind of a step left at the fault trace, as locations of slightly different elevation are brought into juxtaposition. In addition, the rupture itself may not be pure strike-slip; there may be a component of dip-slip motion that helps to raise a little cliff at the fault trace. This cliff, even if it's only an inch or two high, will be obvious in the stratigraphic record as a sharp vertical offset in the layers of deposits.

Let's look at an actual cross section of a fault trace at Pallett Creek. In the photograph, which shows the side of a trench cut by a bulldozer—for Sieh soon tired of working with his shovel—one can see layers of peat and sediment running horizontally in the soil. The approximate dates when some of the layers were laid down (determined by radio-carbon dating) are indicated. In the bottom part of the picture, the layers (including those dated to the years 750, 800, and 1200) are disrupted by the fault trace, which runs diagonally upward from the left-hand corner of the photo toward the upper right. From the layer labeled 1350 onward, however, the layers are not disrupted. Why? Because they were not there when this fault trace was last disrupted by an earthquake—they hadn't been laid down yet. Therefore, the last rupture on this trace took place between about 1200 and 1350. (In fact, the rupture has been dated more precisely to within two decades of 1346.)

The fact that no layers above the 1350 layer are disrupted shows that this particular fault trace has not moved since the earthquake of 1350. But there are other parallel traces, a few yards away, that do carry the record of more recent (as well as older) ruptures. In other words, the "San Andreas fault" is not a single crack in the ground at Pallett Creek, but a set of cracks, some of which ruptured in just one earthquake, others in many earthquakes. To reconstruct the fault's seismic history, it was necessary to put together the data from all the fault traces. According to the final tally, 12 large ruptures have left recognizable traces in the ground. The average interval between events has been 132 years.

One interesting question has to do with the pattern of the earthquakes in time. Right from the start, it appeared that ruptures at Pallett Creek did not occur at regular intervals. But was there some kind of pattern to the timing, or did the ruptures simply occur at random? Sieh's answer to that question has changed several times. The initial dates that he obtained had to be revised more than once, as he got better samples and as the dating techniques became more precise. At first he thought the earthquakes were occurring in closely spaced pairs. Then he thought that there was some kind of cycle in which earthquakes became increasingly frequent and then ceased for a long time. Both of these ideas fell by the wayside as better analyses were made.

In a moment, we'll give you the dates of the Pallett Creek earthquakes in their (we hope) final form, so that you can judge for yourself what kind of temporal pattern there may be. But before doing so, we need to take a diversion and describe another line of research that led to a precise date for the last earthquake before 1857. This research involved the study of tree rings.

Robert Wallace was Sieh's predecessor in this line of research, too. Around 1972 he studied the annual growth rings in trees along the fault

1910

1480

1350

×Youngest disrupted layer

1200

800

750

Wall of an excavation at Pallett Creek showing layers of peat and sediment, some of which are disrupted by slip on one trace of the San Andreas fault (see the text for a further description). (Kerry Sieh and Simon LeVay)

north of San Francisco, looking for evidence of a rupture before 1906. He did in fact find some evidence for an event about 1650. It was equivocal data, and it didn't convince many people, but it now seems he was probably right. In any event, in 1976 Sieh borrowed Wallace's tree corer and took a stroll along the fault where it traverses the San Gabriel Mountains, near Wrightwood and near Frazier Park. He cored quite a few trees, some right on the fault trace, others a good distance away.

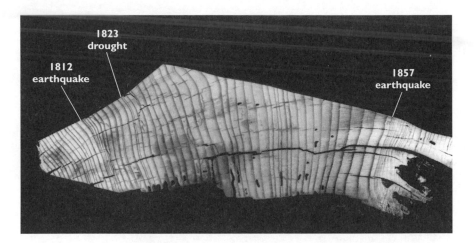

Slab taken from the stump of a white fir on the San Andreas fault in the San Gabriel Mountains. The two zones of thin rings reflect damage from the 1812 and 1857 earthquakes. The narrow ring at 1823 marks a drought that affected trees all over southern California. (Kerry Sieh and Simon LeVay)

When Sieh and his student Kris Meisling examined the cores in the laboratory, they found that some the trees had severe disturbances in their growth patterns, beginning with the ring for the year 1857. Evidently these trees, which stood near the fault, had been traumatized by the earthquake —probably by having major roots severed. This finding confirmed what Sieh already surmised: the 1857 rupture extended beyond Pallett Creek into the eastern San Gabriel Mountains.

Sieh was more interested, though, in using the tree cores to look for evidence of earlier earthquakes. He didn't make any real progress on this objective, however, until he enlisted the help of Gordon Jacoby and Paul Sheppard, dendrochronologists at Columbia University's Lamont-Doherty Geological Observatory in New York. They found that several of the trees had suffered a major trauma in the year 1812. This immediately got Sieh's attention, because 1812 was the Year of the Earthquakes in California. Two large earthquakes occurred near Santa Barbara on December 21, and the San Juan Capistrano earthquake, mentioned at the beginning of Chapter 3, destroyed the church at the San Juan Capistrano mission on December 8.

The Santa Barbara earthquakes were too far away, but the San Juan Capistrano earthquake was another matter. People had always assumed

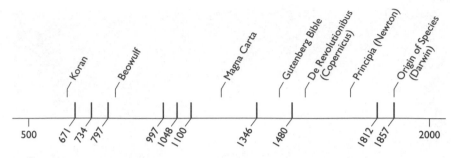

The approximate dates of earthquakes on the San Andreas fault at Pallett Creek, with the dates of some notable Western writings.

that this earthquake had been centered on the coast, near the mission. But now Sieh had reason to believe that it was actually caused by a rupture of the San Andreas fault. As he later found out, the reason the church tower fell down was not that it was shaken particularly strongly but that it was very poorly constructed. The mason who was supposed to oversee the construction had died soon after his arrival, and the work was left to inexperienced padres and Indians.

At this point, then, Sieh had exact dates for two ruptures at Pallett Creek—January 9, 1857, and December 8, 1812—as well as approximate dates for eight previous ruptures, based on radiocarbon analysis. These dates (working backwards) are 1480, 1346, 1100, 1048, 997, 797, 734, and 671. Two earlier earthquakes took place some time between the years 46 and 529.

To give a better idea of the timing of the earthquakes, we have displayed them in graphical form, along with some of the events of historical significance that took place during this same period. What is most striking about the earthquakes is that they seem to have taken place in clusters of two or three closely spaced events, separated by quiescent intervals 200 to 300 years long.

Besides the dates of the ruptures at Pallett Creek, Sieh also wanted to know the amount of slip on each occasion. Finding this out proved to be a formidable task, involving an immense three-dimensional excavation of the fault as well as a study of the warping of the ground on each side of the fault trace. When the work was completed, Sieh had the slip sizes for the last three earthquakes (1480, 1812, and 1857). In each case, they were about 20 feet. These and other data allowed Sieh to determine the long-term slip rate on the Mojave segment of the fault: it was about 35 millimeters (1.4 inches) per year—the same as at Wallace Creek on the Carrizo segment of the fault.

The 20-foot figure was a bit larger than what Sieh had expected, because his initial survey of the 1857 earthquake had suggested that the slip near Pallett Creek was only about 10 to 15 feet. What was really intriguing, however, was how similar the three slips were, even though the ruptures had occurred after quite different intervals of time (about 134 years, about 332 years, and exactly 44 years). It was as if this part of the fault liked to break in ruptures of a constant and characteristic size, regardless of the time available for the accumulation of strain. If strain accumulates at a steady rate (and there is no special reason to doubt that), then Harry Reid's basic assumption—that ruptures of the fault always reset the fault to the zero-strain condition—must not be true, at least at Pallett Creek. The earthquake cycle, originally conceived as a simple clockworklike mechanism, was rapidly becoming more complex and unpredictable.

Wallace Creek—The Sequel

Having figured out the dates of the earthquakes at Pallett Creek, Sieh wondered whether it might be possible to conduct a similar analysis near Wallace Creek, along the Carrizo segment. Several years earlier, he had tried to determine the dates of previous earthquakes there by measuring small streambed offsets. His thought was that streambeds of various ages would be offset by various amounts, depending on how many earthquakes had happened since they were formed. As it turned out, this approach was basically flawed, and it led Sieh to an incorrect conclusion—that the Carrizo segment ruptured in large, regularly spaced earthquakes. He became suspicious of these data when he saw the evidence for clustered earthquakes at Pallett Creek; and so in 1989, Sieh and Lisa Grant (a graduate student) decided to perform an entirely new investigation.

They found a site, 3 miles south of Wallace Creek, where sediment was being laid down on the fault, albeit more sporadically than at Pallett Creek. In this area, which they called the Bidart Fan after the family that leases the land there, Lisa cut a series of trenches and mapped the telltale signs of past ruptures of the fault, using the same techniques Sieh had developed at Pallett Creek.

The results showed that Sieh's earlier analysis had indeed been in error. The 1857 earthquake was confirmed, of course. But before that there was an interval of nearly 400 years without a major rupture. This quiet period was preceded by a cluster of four large ruptures that occurred within a period of about 300 years—between 1200 and 1500. Thus, the Carrizo segment of the fault seems to behave quite like the Mojave segment. The average frequency of earthquakes (about every 160 years) is a bit longer near Wallace Creek, but in both cases the earthquakes group into clusters.

The Big Picture

Besides the excavations at Bidart Fan and Pallett Creek, other sites have also been investigated over the past few years. At Wrightwood, for example, about 16 miles south of Pallett Creek, a group of scientists led by Thomas Fumal and Ray Weldon have excavated a site that contains the record of at least 12 large earthquakes that took place over the last 1300 years. The average interval between earthquakes there has been about 105 years, and as at Pallett Creek, the timing of the earthquakes has been far from regular.

Sieh himself has also studied a site at Indio, in the southernmost portion of the fault north of the Salton Sea. There he found evidence for at least four earthquakes over the past thousand years. Interestingly, the most recent rupture at the Indio site was about 300 years ago, late in the seventeenth century or very early in the eighteenth century. Since the average interval between ruptures at Indio has been about 220 years, the next earthquake there seems to be overdue.

It is tempting to try to assemble the data from the various sites. By figuring out which ruptures at one site correspond with ruptures at other sites, one could hope to determine how far each rupture extended along the fault. This would lead to the complete chronology that Sieh had first dreamed of drawing up back in the 1970s.

In the figure, we have made such an assembly, based on our current state of knowledge. There may have been two ruptures, around 1100 and around 1480, that each broke the entire southern San Andreas fault. If these ruptures did indeed occur as single events, the resulting earthquakes would have had magnitudes of about 8.0—much more powerful than the 1857 earthquake. The dating techniques are

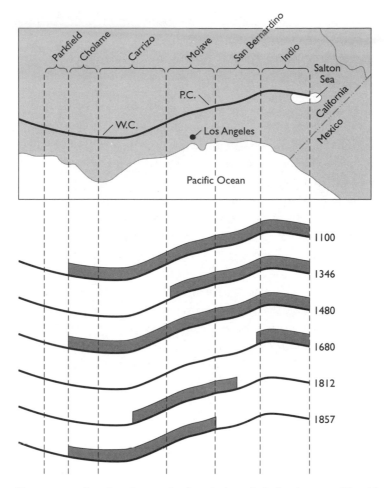

A possible sequence of earthquakes on the San Andreas fault that is compatible with the paleoseismic data from the various excavations. W.C. = Wallace Creek; P.C. = Pallett Creek.

not sufficiently precise, however, to be sure that each rupture was a single event rather than several smaller events within a period of a few decades.

In the period from 1680 to 1857, the southern San Andreas fault broke in a northward-progressing series of three partially overlapping ruptures. The 1906 San Francisco earthquake could be considered a continuation of the northward trend. We cannot say, however, whether this progression is a characteristic feature of the fault's behavior or a

mere chance sequence of events. The earlier history of the fault does
not show such obvious progressions.

The data that we have so far suggest that some aspects of the fault's
behavior have been highly variable. First, the ruptures have varied considerably in length. Thus, the characteristic earthquake model does not
seem to describe the behavior of the southern San Andreas fault very
well. Second, the ruptures at given locations on the fault have not
occurred at regular intervals, and this variability is not explainable by
overlap: for example, some of the ruptures in the Pallett Creek–
Wrightwood area appear to have extended a long way to the north, and
others a long way to the south, but these two subsets, looked at sepa-
rately, still haven't been spaced regularly in time.

Despite this irregularity, there are a few hints of order. First, there
are some indications that the size of slip at a given location may be quite
consistent from earthquake to earthquake, even though the earthquakes
are irregularly spaced in time and involve different lengths of rupture.
In other words, the data we have so far lend some support to the "slip-
patch" model of the earthquake cycle.

Another intriguing suggestion of order concerns the clustering of
ruptures in time. Clustering occurs even in randomly timed sequences
of events. At some locations, however (especially at Bidart Fan and
Pallett Creek), the clustering looks tighter than one would expect on a
random basis, as if something is specifically causing the earthquakes to
bunch together. There aren't enough data to confirm this hunch with a
rigorous statistical analysis, but it is supported by observations in other
parts of the world: clustering of earthquakes has been noted in central
China, the Aleutian Islands, and Turkey, as well as in other parts of the
United States.

If earthquakes really do cluster in some kind of organized fashion,
there are obvious consequences for the prediction of future events. We
might be able to figure out whether a particular segment of the San
Andreas fault is currently within a cluster (in which case the next earthquake could be expected relatively soon) or between clusters (in which
case it would be a relatively long time until the next earthquake). An
ongoing paleoseismic study of the fault by a team of geologists at the
Southern California Earthquake Center is aimed at clarifying whether
we are dealing with random temporal irregularities or with meaningful
clusters.

We'll come back to these two apparent kinds of order (slip patches and clustering in time) in the next chapter, after we have taken a look at how scientists, using a variety of new and old techniques, are monitoring the fault's current activity. By combining the historical record with an understanding of the fault's dynamic behavior, we can gain more insight into the earthquake cycle and make more informed statements about the probabilities of future ruptures.

5

LYING IN WAIT: THE SAN ANDREAS FAULT TODAY AND TOMORROW

If ever there was a place that time forgot, it is Parkfield, California. From Coalinga, itself a sleepy enough town in the San Joaquin Valley, it's only 20 miles south-southwest as the crow flies, but the dirt road that winds up and over the Diablo Range makes it seem like double or triple the distance. As you negotiate the endless off-camber turns, wait for cows to trundle out of your path, or pause to take in the ever-changing vistas, your whole notion of time begins to relax and spread out. By the time you come down through the oak-shaded valley, past the ostrich ranch, and into the tiny village, you are probably an hour behind schedule. But schedules no longer matter.

Oak tree growing directly on the creeping segment of the San Andreas fault. The trunk has been split, with the near side moving to the left and the far side moving to the right, by about the length of the student's hand. (Kerry Sieh and Simon LeVay)

Parkfielders (all 37 of them) live in geological time. In 31.5 million years, according to a sign next to the Parkfield Inn, Parkfield will be a seaside community. The sign itself has been cleft in two, and the halves now stand 12 feet apart. No, the San Andreas fault didn't do that. It's just artistic license, a faux-seismological depiction of Parkfield's manifest destiny. "Earthquake Capital of the World," announces another sign across the street, hand-painted on a boiler-turned-weed-planter, "Be Here When It Happens." It will be a bumpy ride to the beach.

Parkfield is the southern sentinel of a 100-mile stretch of the San Andreas fault whose behavior is quite unlike the more notorious reaches to the north and south. These neighboring segments, which broke in the great earthquakes of 1906 and 1857, now lie locked in seismic silence—a silence as profound as that of Cascadia's subduction zone. But the segment from Parkfield north to San Juan Bautista is a seismic chatterbox. On practically any given day you can pick up an earthquake or two, if your instruments are sensitive enough. Not big ones, for sure. Small ones, tiny ones, microscopic ones. By little fits and starts, and by motions too slow and steady to register on seismographs at all, the two sides of the fault are slipping past each other, night and day, year in and year out.

Why the "creeping segment" behaves the way it does is a mystery, although as with most mysteries there are plenty of hypotheses. Perhaps the stresses that hold the two sides of the fault clamped together are weaker here. Perhaps there is more water in the rocks, making this part of the fault more slippery. Perhaps there is something unusual about the detailed geometry of the fault surface or the mineral makeup of the rocks that are in contact. Until someone drills down a few miles and takes a close look, it will remain a matter of speculation.

The creeping segment is a geological honey pot: it attracts seismologists from far and wide. For seismologists have a strangely perverse view of earthquakes—that there aren't nearly enough of them. Earthquakes are what sustain a seismologist's professional life and give it direction; it is the intervals between them that are deadly. But along the creeping segment, there are earthquakes aplenty, and the intervals are barely noticeable.

Bill Ellsworth, a seismologist at the USGS in Menlo Park, is one devotee of the creeping segment. "Theories of earthquakes are very difficult to test," he says, "because we have such a short history of earthquakes. So what can we do to try to understand the process? One

approach is to look at the small earthquakes, simply because they happen more frequently. If we can look in detail at the behavior of a small
region where lots of small earthquakes are happening, we can hope to
understand which of the various models of the earthquake cycle is the
best description of what's actually going on. And the answer seems to
be, at least on the creeping part of the fault, that the 'characteristic
earthquake' model is the right one: most of the earthquakes seem to
show a repetitive behavior, where an earthquake that occurs at a particular place has also occurred many times in the past, with little or no
variation."

Ellsworth's conclusion is based almost entirely on close analysis of
seismographic records. Seismologists long ago figured out the basic
techniques for using these records to determine an earthquake's vital
statistics: the location of its source, the orientation of the fault plane, the
approximate size of the region of the fault that has ruptured, and the
amount and direction of slip. To locate the source, for example, seismologists measure the arrival time of two different kinds of seismic
waves, the fast-moving compressive or "primary" waves and the slower
moving transverse or "secondary" waves. Although the two kinds of
waves leave the source at the same time, the primary waves gradually
get farther and farther ahead of the secondary waves. Just as one can
estimate the distance of a lightning strike by counting the seconds
between the arrival of the fast-moving flash of light and the slow-mov-
ing clap of thunder, so one can calculate the distance of an earthquake
source by measuring the time between the arrival of the primary and
secondary waves at the seismographic station. This measurement puts
the source somewhere on a sphere whose radius corresponds to the calculated distance. If the records from four different stations are analyzed
in the same way, the source, in principle, must be located at the unique
point where the four spheres intersect. In practice, however, there's a
certain fuzziness to the result, because it is limited by the accuracy with
which one can measure the wave fronts, as well as by ignorance of the
density of the intervening rock and hence of the exact speed of wave
travel. The resulting uncertainty can easily extend to hundreds of yards,
and sometimes a mile or more. Similar uncertainties affect the estima-
tion of the area of the fault that ruptured and the amount and direction
of slip.

But when Ellsworth pored over the records of hundreds of small
earthquakes that had emanated from a particular region of the creeping

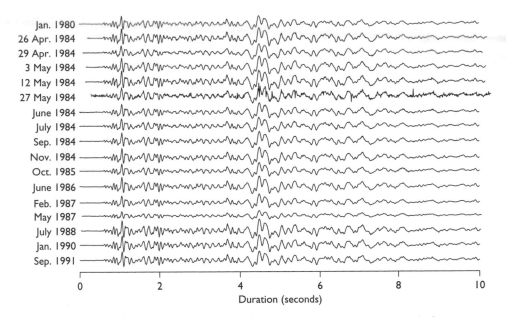

Seismographic traces showing 17 almost identical earthquakes occurring over a period of 12 years. (Courtesy of Bill Ellsworth)

segment over a period of 20 years or so, he made a remarkable observation. Some of the records duplicated each other, almost down to the last little squiggle. In fact, the majority of all the earthquakes Ellsworth examined were repeating earthquakes of this kind—they were clones of previous earthquakes, and they were followed by clones of themselves. A given short stretch of the fault—say, a couple of miles long—might give rise to 10 or 20 different sets of repeating earthquakes, with each set consisting of a dozen or more individual earthquakes over a 20-year period of observation.

The similarity of the records within a set suggested that the location and size of the patch of the fault that broke in one earthquake were similar to the location and size of the patch that broke in the next earthquake. How similar, Ellsworth was able to determine with a more sophisticated analysis. He compared the various members of a repeating set using a technique called cross-correlation analysis, which uses information from the entire waveforms, not just from the onset or termination of the various kinds of waves. With this technique, he could determine the time of arrival of the earthquake signals at the recording stations to within a few milliseconds. He could then compare the loca-

tions of the successive earthquakes within a repeating set, with a precision perhaps a hundred times better than would be possible with the
standard procedure. He found that if one earthquake involved the rupture of a patch measuring, say, a few hundred feet across, other members of the set would rupture a patch that was identical to within an
accuracy of a few feet. It was clearly the same earthquake repeating over
and over.

So there seem to be small, isolated patches of the fault that are more
securely locked than the surrounding parts of the fault plane. The surrounding areas creep steadily, but the patches rupture intermittently, and
the successive ruptures of a single patch produce characteristic earthquakes. One might think that after a few earthquakes, the two sides of
the patch would have slid past each other, so that the patch as a coherent entity would no longer exist. But in fact, the amount of slip per
earthquake is much smaller than the size of the patch (that is, inches of
slip on a patch a few hundred feet across), and so it may take thousands
of years for a patch to be completely separated into its two halves.

The timing of the earthquakes within a repeating set is considerably more variable than the location and size of the rupture. Some
patches, especially those that are well isolated from others, break at
nearly regular intervals. Other patches, including those that happen to
be grouped into clusters of several adjacent patches, may break at quite
irregular intervals. Ellsworth believes the crucial factor that determines
when a patch fails is the amount of strain that has built up. If the patch
is well isolated, strain may accumulate very uniformly over time, leading to very regular earthquakes; but if a patch is near other patches, or
if a much larger earthquake somewhere in the region increases or
decreases the strain on the patch, the timing will be altered. What does
not change is the tendency for discrete patches to break in ruptures of
constant size.

The striking thing about Ellsworth's results is that they could so
easily have been completely different. Quite a few seismologists would
have expected the behavior of the creeping segment to be describable
only in statistical terms—in terms of the probabilities of ruptures within certain regions and certain ranges of size over certain periods of
time. If this were true, no one earthquake would resemble any other.
But instead, the earthquakes are like old friends: you may not see them
for a while, but when you do, you recognize them instantly. Ellsworth's
dream is to get to know them even more intimately: he wants to drill

down to one of his patches and see exactly what is special about it that gives it its characteristic behavior. Even the shallowest of the patches, however, is a couple of miles down, and it is a small enough target to challenge the skill of the world's best drillers. So for starters, Ellsworth wants to drill about a mile down, to see what may be causing the fault to creep: whether high fluid pressure, exotic rock chemistry, or something else. Later, perhaps, it may be feasible to extend the drilling to hit one of the patches.

The Parkfield Earthquake

Can one generalize from Ellsworth's observations of small earthquakes to moderate and large earthquakes? A critic could argue that these small earthquakes are not typical, because they occur on "stuck" patches that are embedded in a gently creeping matrix. Perhaps everything changes as the scale of things increases. After all, we already noted in Chapter 4 that, if the paleoseismological record is to be believed, very large earthquakes on the San Andreas fault have not always mimicked previous earthquakes. So are there characteristic earthquakes of a size that do damage?

Parkfielders will tell you that there are. They have one. It's a magnitude 5.6 jolt that has been plaguing them since at least the middle of the last century. It revisits them every 22 years or so, like a ghost that refuses to be exorcised, knocking down chimneys, overturning furniture, and ripping up tarmac on the village street.

The last time the earthquake came visiting was on June 27, 1966, in the middle of the evening. It announced its arrival with a foreshock, magnitude 5.1. Then, 17 minutes later, the mainshock dumped the contents of Donalee Thomason's china cabinet onto her parlor floor. Because she was warned by the foreshock, Donalee had already gotten her most precious items into the safety of a Styrofoam cooler.

Donalee knew the first shock was a harbinger of something bigger, because she remembered the previous time, back in 1934, when she was nine years old. Then, too, there was a 17-minute pause between the two shocks, although the villagers, who were gathered for a school play in the community center, didn't take advantage of the warning to leave the building. When the mainshock came, there was a lot of screaming and rushing for the exits, but no one was seriously hurt.

The time before that was in 1922; before that, 1901; before that, 1881; and before that, 1857. Six earthquakes, all pretty much alike, at an average interval of 22 years. And if you're willing to say that the 1934 event "came early," then you can make the case that there's some under-lying clockworklike mechanism that sticks to the 22-year cycle quite closely. At least, you could have said so until 1988, when the next earth-quake was due. It's now 10 years later and counting, and still no earth-quake, much to the chagrin of a legion of seismologists who have out-fitted the village and its environs with seismographs, accelerometers, dilatometers, laser range finders, creep meters, magnetometers, well gauges, and a host of other low- and high-tech instruments, all with the intent of catching the reluctant earthquake in the act.

Even before its next visit, however, there is plenty of reason to call the Parkfield earthquake a characteristic earthquake. Again, the evi-dence comes mainly from the seismographic records. Ellsworth and his colleagues have made a close comparison of the records from the 1934 and 1966 events. On both occasions, the foreshock had a magnitude of 5.1, and on both occasions it was centered at a point 5.5 miles below the surface and about 7 miles northwest of the village. The waves from the 1934 and 1966 foreshocks—even the faint reverberations a minute or so after the beginning of the shock—can be superimposed precisely on top of each other: the two events were identical. The length of time between the foreshock and the mainshock was almost identical on the two occasions: 17 minutes and 25 seconds in 1934; 17 minutes and 17 seconds in 1966.

With the mainshock, it's a little more complex. The focus (or point of origin) of the mainshock was certainly located at the same point on the two occasions: it was about a mile closer to the village than the fore-shock and a little nearer the surface. But although the initial portions of the two mainshock traces can be overlain on each other, they begin to differ later in the event. "The two earthquakes probably grew to differ-ent lengths," says Ellsworth. "Thirty-four seems to have been a shorter rupture than '66. And I think that's an important observation in terms of applying the characteristic-earthquake model to long segments. Earthquakes may begin in a characteristic way: once that patch of a par-ticular size fails it produces a characteristic earthquake, but what we can't say is whether the next patch will also fall down as the next domino."

One intriguing question is: What is happening during the 17 min-utes between the Parkfield foreshock and the mainshock? For the interval to repeat so closely, something in the Earth must be keeping track of the time. No one knows for sure what that is, but one possibility is that the minutes are counted off by some kind of uniform sliding of the rock faces against each other—a sliding that is much faster than the background creeping motion of the fault, but still much too slow to register on seismographs. Perhaps a moving front of this accelerated creep migrates from the focus of the foreshock toward the focus of the impending mainshock, and it takes 17 minutes to traverse the distance.

Episodes of accelerated creep, or "creep events," have in fact been detected on a number of occasions. A particularly dramatic example occurred on the San Andreas fault in December 1992, at the northern end of the creeping segment near San Juan Bautista. On that occasion, 11 square miles of the subterranean fault surface slid about an inch over a period of 7 to 10 days. The event did not generate any seismic waves, but it was detected by strain meters that had been installed near the fault previously. Alan Linde, the Carnegie Institution scientist who made the discovery, calls the event a "slow earthquake." On the moment magnitude scale, it would merit a 4.8—enough to cause a sizable jolt if it had been compressed into a few seconds. Furthermore, the event did cause a series of aftershocks of the conventional kind; the largest was a 3.7.

Although the 1992 event was not associated with large conventional earthquakes, there have been examples in which "creep events" or slow earthquakes *have* preceded or followed conventional earthquakes. The giant Chilean earthquake of 1960, for example, was preceded by a slow earthquake, and a magnitude 7.0 earthquake near Tokyo in 1978 was followed by a slow earthquake that lasted two hours. Some scientists who study the fracturing of rock in the laboratory, such as Jim Dieterich of the USGS, Menlo Park, believe that creeplike episodes may be necessary precursors to the catastrophic failures that generate conventional earthquakes. Thus, although slow earthquakes are difficult to detect, they may play a crucial role in triggering earthquakes and linking one earthquake with another.

The Parkfield mainshock is only a moderate-sized earthquake, and it is worth considering whether this earthquake might, in the right circumstances, play the role of foreshock to an even larger rupture. In fact, that may be what happened in 1857. There were two foreshocks, each about a magnitude 6 and each centered somewhere in the Parkfield area, followed a few hours later by the great Fort Tejon mainshock.

If Parkfield did trigger the Fort Tejon earthquake, could it do the same on its next appearance, an appearance that could already be considered a decade overdue? Presumably, that would depend on whether the Carrizo segment, south of Parkfield, has reached a level of strain that could sustain a rupture. As we discussed in Chapter 4, the Carrizo segment ruptures about every 160 years, on the average, with quite a bit of variability in timing. So it is well within the bounds of possibility that this segment is now ready to break and needs only a minor jolt from the north to do so. If this scenario comes to pass, the Parkfield rupture could play foreshock to a Carrizo rupture, preceding it by a few minutes or hours, or else the Parkfield rupture could simply propagate non-stop through the boundary between the segments, giving rise to a single, very large earthquake.

Of Dominoes and Modules

This train of thought leads to a revised version of the characteristic-earthquake theory. For small earthquakes, caused by rupture of small patches of a fault, the ruptures may be truly characteristic, breaking again and again in near identical fashion. But larger ruptures may be thought of as modular, being assembled from a set of smaller patches that on this occasion rupture together but are not necessarily destined to do so on every occasion.

A good example of such modular organization comes from the Imperial fault, a 37-mile-long strike-slip fault that runs south from the Salton Sea, across the international border, and about 17 miles into Mexico. The Imperial fault is a southward continuation of the San Andreas system; alternatively, it can be thought of as the northernmost transform fault of the spreading center that zigzags its way up the Gulf of California (see Chapter 3).

There have been two ruptures of the Imperial fault during the period of instrumental surveillance: the first in 1940 (the El Centro earthquake) and the second in 1979 (the Imperial Valley earthquake). Evidently these were not characteristic earthquakes, because the 1940 event had a magnitude of 7.1 and was accompanied by surface offsets of up to 20 feet, whereas the 1979 event had a magnitude of 6.6 and was accompanied by a maximum slip at the surface of only about 2 feet.

Rows in a newly planted onion field, offset by about 6 inches during the 1979 Imperial Valley earthquake. (Kerry Sieh and Simon LeVay)

Citrus trees offset by about 19 feet during the 1940 El Centro earthquake. (Courtesy of Clarence Allen)

High-altitude oblique aerial photograph of the Imperial and Mexicali valleys showing the
trace of the Imperial fault. The dotted portion ruptured in the 1940 earthquake; the solid
portion ruptured during both the 1940 and 1979 earthquakes. The square fields are
about a mile on a side. (Government photo, courtesy of Robert Sharp, USGS)

The interesting features of these two earthquakes become apparent,
however, when one maps the amount of slip along the length of the fault
on the two occasions. The 1979 earthquake was caused by the rupture of
only a northern segment of the fault, extending from the Salton Sea to
within about 4 miles of the U.S.-Mexican border, where it terminated
rather abruptly. The 1940 earthquake was caused by rupture of both this
northern segment and the remaining, southern portion of the fault, but
the amount of slip was very different in these two regions. The southern
portion slipped by up to 20 feet, but the northern portion slipped by only
about 2 feet: in fact, the principal aspects of the slip along this segment
were quite similar to what was observed in the 1979 rupture.

These observations lead to the idea that the Imperial fault has a
modular organization, composed of at least two segments that each rup-

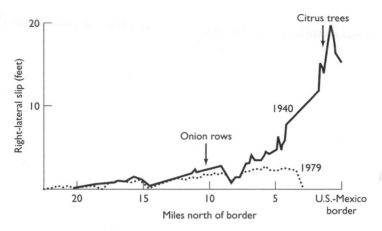

Comparison of the slip on the Imperial fault in the 1940 and 1979 earthquakes. (Adapted from an illustration by R. Sharp, USGS)

ture in a characteristic mode: a northern segment that breaks relatively frequently with a small slip, and a southern segment that breaks relatively infrequently with a large slip. The 1940 event, according to this idea, was an assembly of both modules, whereas the 1979 event involved only the northern module. (The southern portion itself may comprise two segments, for there is a paleoseismic hint that the southernmost 12 miles of the fault break in smaller, more frequent ruptures than does the central segment, near the U.S.-Mexican border.)

Seismographic records tell us that the 1940 event began in the northern segment and propagated through the segment boundary, triggering the southern segment to rupture also, like one domino knocking down its neighbor. By 1979, the northern segment had accumulated enough strain to rupture again, but the southern segment, which seems to lock more tightly, was not ready to rupture. The 1979 rupture began at the base of the southern segment, but sustained no large slippage until the rupture front advanced northward into the northern module. Apparently, the southern module was not ready to slip again.

The notion of large ruptures as a chain of smaller, characteristic ruptures that trigger one another is a little too simplistic. In fact, neighboring segments can influence each other to break in an uncharacteristic manner. We can see that by looking more closely at the slip in the 1940 earthquake. If we assume that the boundary between the northern and southern segments of the Imperial fault lies where the 1979 rupture terminated, 4 miles north of the U.S.–Mexican border, then we can see that the 1940 rupture caused the southernmost 5 miles or so of the northern segment to break by more than its characteristic slip—this region slipped more in 1940 than it did in 1979. Evidently, this region was induced to slip uncharacteristically far in 1940 by the much larger slip close by on the southern segment. The segment boundary is evidently a step up in strength as one goes southward, or a step *down* in strength as one goes northward. Thus, stress changes to the south of the boundary can easily affect what happens to the north, but stress changes to the north cannot easily affect what happens to the south.

The same thing is probably true back in Parkfield. We described earlier the small patches that break in highly characteristic small earthquakes. Some of these patches, however, are within the larger zone that ruptures in the 22-year Parkfield earthquakes. What happens to the small patches during these larger ruptures? In all probability, they go with the flow, slipping by the same amount as the entire rupture surface. In other words, the major stress changes associated with the larger rupture cause the small patches to abandon their characteristic behavior for the day and to slip by many inches rather than an inch or less. Later, they resume their characteristic behavior.

So, we are beginning to develop a notion of large earthquakes as being caused by the combined rupture of a number of segments, modules, or patches—call them what you will. Each segment, left to itself, likes to break in a characteristic fashion, but may break uncharacteristically under the influence of stress changes that accompany the rupture of neighboring segments. On some faults, the same assemblage of subunits tends to break together on repeated occasions, producing truly characteristic large earthquakes. More commonly, though, successive earthquakes on a fault involve somewhat different assemblies of segments. Forecasting earthquakes then becomes a question of predicting which assembly of segments will break together in the next quake and when at least one segment within this assembly will be stressed to failure, triggering the entire assembly to rupture.

Stresses and Shadows

What, then, determines when a segment of a fault ruptures? The correct (but facile) answer to this question is that it ruptures when the forces tending to make it rupture overcome the forces tending to prevent it from rupturing. Let's try to give this a little more meaning by visualizing the actual forces that are work.

Imagine two metal slabs (representing the two sides of the fault) held together by a vise. Two people grab and pull on opposite ends of the two slabs, trying to slide them past each other. A third person turns the handle of the vise, trying to press the slabs so tightly against each other that they resist sliding. We can see in a general way that, if the slabs are to slip against each other, the force generated by the two people tugging (a force called the *shear stress*) must exceed the force exerted by the vise (we'll call it the *clamping stress,* although technically it's known as the *normal stress*).

There's an extra complication. Let's imagine that a fourth person drops oil from an oilcan onto the line of contact between the two slabs. When this oil penetrates between the slabs, it's going to make them a lot more slippery, and the force exerted by the vise may no longer be sufficient to prevent the slabs from sliding against each other. In other words, the effectiveness of the clamping stress in resisting rupture is modified by the properties of the fault surface: its roughness, the presence of fluid in the rock, and the pressure exerted by that fluid—properties that could be lumped together as a "slipperiness factor." In the extreme situation, the clamping stress may become irrelevant. Think of a block of ice on a warm tabletop: the layer of meltwater under the ice allows it to slide freely, even if an elephant climbs on top of it.

In the case of a strike-slip fault such as the San Andreas, the shear stress is mainly the force exerted by the moving tectonic plates, a force that is communicated to the walls of the fault from below and from the sides. In Reid's original conception, described in Chapter 3, the shear stress is reset to zero by a rupture—the fault is fully "relaxed"—and then steadily increases over time. We now believe that fault ruptures don't reset the shear stress to zero, and in fact we have no easy way to determine what the shear stress on a given fault may be. (We can measure the stress near the surface, but this is not the same as the stress miles underground.) We can calculate, however, the *rate* at which tectonic forces are adding stress to a fault. In the case of a locked segment of the San Andreas fault, for example, plate motion adds about a fifth of a bar of shear stress to the fault every year. (One bar is roughly equivalent to the force per unit area exerted by the Earth's atmosphere at sea level.) In addition to the gradual increase in shear stress caused by plate motion, ruptures of neighboring segments of the fault or of other faults in the vicinity may cause sudden, steplike increases or decreases in the shear stress.

The clamping stress on a fault comes primarily from the weight of rock lying above a given point on the fault plane—the *load stress*. If a fault, such as the San Andreas, is roughly parallel to the direction of relative motion between the tectonic plates, then there will be no systematic increase or decrease of the clamping stress during the interval between earthquakes. As we will see in a moment, however, the clamping stress can be changed by ruptures of other nearby faults. These changes can be calculated, and when combined with an estimate for the slipperiness factor, they tell us how much the force opposing a rupture has increased or decreased.

A group of geophysicists at Menlo Park, including Ellsworth, Ross Stein, Ruth Harris, Paul Reasenberg, and Robert Simpson, have used this simple model to calculate how a given earthquake affects the likelihood of future earthquakes on other faults in the vicinity. They started with the 1906 San Francisco earthquake. Knowing the amount of slip that occurred along the fault on that occasion, and treating the Bay Area as a uniform slab of rock of known elasticity, the scientists calculated how stress changes should have been distributed throughout the area. Looking at the Hayward fault, for example, a right-lateral strike-slip fault that runs parallel to the San Andreas fault on the east side of San Francisco Bay, the group found that the shear stress was decreased by the San Francisco earthquake, and the clamping stress was more or less unaffected. Thus, the next rupture of the Hayward fault was postponed by the San Francisco earthquake. In fact, since most of the faults in the Bay Area are parallel right-lateral faults, almost all of them were "relieved" by the 1906 earthquake. For this reason, the Menlo Park scientists speak of a "stress shadow" that was cast over the entire Bay Area by the 1906 earthquake, reducing the probability of other earthquakes on most faults.

In fact, the 1906 earthquake does seem to have caused a drop in the frequency of moderate-to-large earthquakes in the Bay Area. In the 70 years *before* 1906, there had been 16 earthquakes of magnitude 6 or greater; in the 70 years *after* 1906, there was only one such earthquake. But in the late 1970s, earthquake activity began increasing in the southern part of the Bay Area. The stress shadow is getting smaller, the scientists believe, as the inexorable motions of the tectonic plates begin to rebuild the stress patterns that had existed before 1906.

The Menlo Park group (specifically, Harris and Simpson) has done a similar analysis for the 1857 Fort Tejon earthquake, using Sieh's measurements of the slip in that earthquake as a starting point. The faults in southern California are much more diverse in their orientation and in the direction of slip that occurs on them. Thus, Harris and Simpson found that the 1857 earthquake should have caused the rupture of some faults in the vicinity to be postponed, whereas the rupture of other faults should have been brought closer. And of the 13 moderate to large earthquakes that occurred in southern California during the 50 years after 1857, at least 11 and perhaps all 13 did in fact occur on faults where, according to Harris and Simpson's calculations, ruptures should have been brought closer. In other words, their analysis of stress redistribution really seems to say something useful about the likelihood of earthquakes in different locations during the 50 years after 1857. After 1907, however, earthquakes began occurring indiscriminately in southern California, suggesting that the effects of the 1857 rupture had been erased by the gradually increasing stresses coming from the plate motion.

As Hiroo Kanamori, former director of Caltech's Seismological Laboratory, has noted, "Forecasting earthquakes is very difficult—especially ahead of time." The crucial test of the Menlo Park group's approach will come when it attempts to predict future earthquakes.

The group members have made a start in this direction by analyzing the expected effects of an earthquake that occurred near the small town of Landers, north of Palm Springs, in 1992. The Landers earthquake had a magnitude of 7.3, making it the largest to strike California in 40 years. Furthermore, it was exceptionally well documented—by seismographic recordings, by GPS measurements of deformation, and by precise measurements of slip along the rupture. These measurements were critical for Harris and Simpson, because to do their calculation they needed accurate information about the changes produced in the ground by the earthquake. One new technique has confirmed the validity of the models based on these traditional methods. Called *space-based synthetic-aperture radar interferometry,* the technique is capable of mapping the deformation of the land surface for many miles around a fault. A satellite makes a radar scan of the region before and after the rupture. (If no scan was taken before the rupture, one is simply out of luck). The two scans, which encode the precise distance between the satellite and the ground, are then digitally subtracted one from the other, leaving an image that represents (in the form of colored bands called interference fringes) the amount by which the ground has moved as a consequence of the earthquake. With this technique, elastic deformation of the ground could be detected as far as several tens of miles away from the Landers rupture (see Color Plate 10).

When Harris and Simpson (and two other groups) applied their analysis to the Landers earthquake, they got a complex pattern of increased and decreased stresses on other faults in southern California (see Color Plate 11). Shear stresses on two faults that flank the Landers rupture (the Calico and Lenwood faults) were relieved, meaning that the next ruptures on these faults, if the analysis is correct, have been postponed for many years. Parts of the San Andreas fault were also relieved, although the changes were small. Along a segment of the San Andreas fault near San Bernardino, however, the shear stresses were increased by up to 6 bars; in addition, the clamping stress was reduced. These stress changes, according to Harris and Simpson, are about 14 times greater than the annual stress changes caused by tectonic motion. Therefore, it appears that the next rupture of this segment—a rupture that is expected to produce a great, magnitude 7.5–8.0 earthquake—has been brought 14 years closer by the Landers earthquake.

Of course, the limitation of this kind of forecast is that we don't know when the San Bernardino segment was due to rupture in the absence of the Landers earthquake. If the segment was due to rupture 15 years from now, a 14-year advance would be sinister in the extreme; if 100 years from now, it would be relatively unimportant. In that sense, the stress *shadows*, which put off upcoming earthquakes by a certain number of years, are more informative than the stress *increases*, which bring us closer to earthquakes whose dates nevertheless remain very uncertain.

The Earthquake Synthesizer

If interactions between neighboring segments of a fault, and between neighboring faults, greatly influence when a segment ruptures, the tendency of small patches to break characteristically may easily become submerged in a more chaotic pattern of large earthquakes. Steve Ward, a geophysicist at the University of California, Santa Cruz, has explored this possibility by means of computer models—specifically, by developing programs that create long sequences of artificial earthquakes.

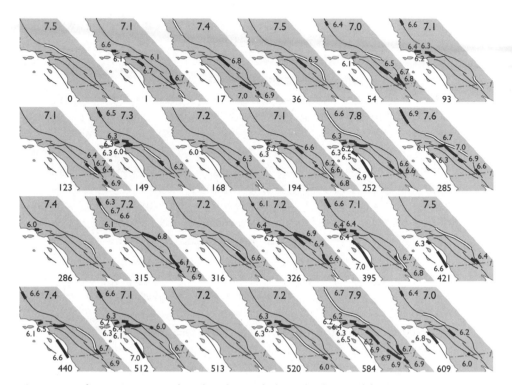

A sequence of computer-generated earthquakes on faults in southern California, over a span of 609 years (the year is indicated at the bottom of each frame). Each frame shows the location of a fault rupture causing a magnitude 7.0 or greater earthquake (white line), as well as all ruptures causing magnitude 6.0–6.9 earthquakes that have occurred since the previous frame (black lines).(Courtesy of Steve Ward)

One of Ward's models generates artificial earthquakes on 14 major faults in southern California, including the San Andreas fault. As input to the model, he uses the vital statistics for these faults, which are broken down into a total of 39 individual segments. These statistics include the location, length, overall slip rate, characteristic slip per earthquake, and estimated fault strength for each segment. He also gives the computer information about the elasticity of rock as well as other physical properties that govern how stress is transferred from one point in the Earth to another. Then he starts the program, stands well back, and allows the computer to fire off a few millennia's worth of earthquakes.

When the digital dust has settled, Ward is left with an "earthquake movie"—a frame-by-frame depiction of the major fault ruptures in southern California over thousands of years. In the movie, one can see an intriguing mixture of order and disorder. The Carrizo segment of the San Andreas fault, a very strong, rupture-resistant segment, breaks fairly regularly in large-magnitude earthquakes with uniformly large slips. (In fact, the model may have this segment behaving more regularly than the most recent paleoseismological data, discussed in Chapter 4, would suggest.) The regularity reflects the fact that, in the model, stresses transferred from other fault segments are not capable of greatly advancing or retarding the Carrizo segment's ruptures; their timing is therefore controlled primarily by the plate motion. Conversely, the rupture of the Carrizo segment does have a great impact on other segments, sometimes causing neighboring segments to rupture in the same event, at other times causing them to rupture in separate earthquakes a few years later. Furthermore, ruptures of the strong segments such as Carrizo leave pronounced stress shadows over large areas, so that the overall frequency of earthquakes oscillates over periods of many decades.

According to the model, relatively weak segments such as the Mojave segment (where Pallett Creek is located) are much less regular in their behavior. Although the Mojave segment has its own characteristic earthquake—a magnitude 7.5 event with a slip of about 15 feet—these earthquakes do not occur at all regularly. Furthermore, they are interspersed with other, uncharacteristic ruptures; these are usually occasions when the Mojave segment slips by a relatively small amount as part of a longer rupture involving segments to the north or south.

Ward's model tends towards chaotic behavior. Although everything that happens is deterministic—that is, it happens as a consequence of measurable causes—very small differences in initial conditions can have radical differences in results. For example, a slightly greater stress transfer from one fault to another may change the timing and mode of rupture of the second fault, which in turn changes everything down the line. Since we will never know the strengths of these interactions with 100 percent precision, we may never be able to make precise long-range forecasts of earthquakes in a complex system such as the one modeled in Ward's computer.

But long-range in this context means through several cycles of earthquakes. In principle, a model such as Ward's is very amenable to short-term prediction; that is, simply figuring out where and when the next earthquake or two will happen. In fact, Ward proposes using his model for exactly that purpose. It would work as follows. A neural net—a computer program that uses what are thought to be brainlike learning processes—would be exposed to thousands or millions of years of the earthquake movie, until it had figured out how to predict the location and size of one or two upcoming earthquakes. Then it would be let loose on the actual history of earthquakes in southern California over the past hundred years or so and asked to tell us what will happen next.

Almost certainly, it wouldn't work, because there are probably all kinds of errors in the information used to construct Ward's model. Neural nets are generally trained on *real* data; training them on synthetic data would probably cause them to develop serious misconceptions. Ward himself misses no opportunity to stress the limitations of the model—it doesn't even include the faults that produced the Landers earthquake, for example—or to point out that the model can be no better than the data used to construct it.

What he does claim, however, is that his modeling approach is likely to do as well as or better than any other approach, because it tries to take into account most of the factors that are believed to influence the occurrence of earthquakes. Although the model might not be the basis of a useful prediction right now, it will undoubtedly be refined as more real earthquakes happen and the manner in which faults interact becomes better understood. In fact, it is hard to see how any approach other than computer modeling, grounded in sound and abundant data, can hope to provide accurate predictions in the complex system formed by the San Andreas fault and its neighbors.

The really dreaded outcome of this line of research, of course, is a computer program that can tell us where and when the next earthquake will occur, but can't tell us, in terms we can understand, *how* it figured it out. At that point—still a long way off, we hope—scientists will be reduced to machine-tending technicians.

All this attention paid to the San Andreas fault, over the past three chapters, might lead one to believe that it is the only significant strike-slip fault in the world. Far from it. Although the San Andreas fault is the most famous example, many others present significant hazards. In the United States, the other major strike-slip fault is the Denali fault, which runs for 300 miles across the glaciated mountains of southern Alaska. It hasn't broken in historical times, and it slips at a long-term rate only one-third that of the San Andreas fault, but the geological record suggests that it slipped by up to 45 feet during its most recent rupture. An event of that size must have caused a very large earthquake indeed. The Denali fault does not run along a discrete plate boundary, and its tectonic mechanism is not well understood.

The 500-mile-long Alpine fault traverses the mountainous western side of New Zealand's South Island and marks the line along which the Australian and Pacific plates are sliding against each other. Rather like the San Andreas fault near Los Angeles, the Alpine fault is oriented at an oblique angle to the direction of plate motion. This obliquity causes a compression, which has thrown up the magnificent Southern Alps. From geological evidence, it is clear that the Alpine fault has ruptured repeatedly over the past few hundred years. It is probably every bit the equal of the San Andreas fault in terms of its potential for unleashing great earthquakes; it just doesn't happen to have done so in the 160 years since Europeans settled in New Zealand.

Another strike-slip fault zone, the Dead Sea transform fault, runs north–south through the Dead Sea, where the Arabian Plate is moving northward relative to the African Plate (thus, it's a left-lateral fault, the opposite of the San Andreas fault). Some seismologists and archeologists believe that it was an earthquake on this fault about 3000 years ago, and not the blowing of seven ram's-horn trumpets, that brought down the walls of Jericho. In fact, the city has been damaged or destroyed by earthquakes at least a dozen times over its history. Farther north along the fault, where it traverses Lebanon's Beka'a Valley, the fault appears to have generated large earthquakes about every 500 years, most recently in 1202 and 1759.

This brings our tour of the San Andreas fault to a conclusion. Our next stop, though, is not far afield. For one seemingly minor feature of the fault, the bend between the Carrizo and Mojave segments, has had fateful consequences for Los Angeles. It has spawned another set of faults, quite different in their mode of action, right under the metropolis. Such urban faults, though they may not deliver the great earthquakes for which the San Andreas fault is famous, more than make up for that by their proximity to vulnerable structures, as Los Angeles found out in January 1994.

6

THE ENEMY WITHIN THE GATES: EARTHQUAKES ON URBAN FAULTS

It was a rude awakening: 10 million residents of southern California were torn from their private slumbers to share a half-minute of communal terror. Eleven miles below the town of Northridge, in Los Angeles's San Fernando Valley, rock stressed to the breaking point had suddenly failed. A 10-mile-wide crack raced upward at 7000 miles per hour, radiating enough energy to bring down freeways; knock houses off their foundations; and rupture gas, water, and power lines. For each person who went through it, the Northridge earthquake of January 17, 1994, was an indelible personal experience, but it was also an event that

united the entire community in an effort to surmount the psychological and economic damage.

Both of us were awakened by the earthquake. Simon LeVay and his
partner Mike were asleep on the top floor of a wood-frame apartment
building in West Hollywood, about 13 miles southeast of Northridge.
When they awoke, the building was being pitched alternately northward and southward about once every second. Each change of direction was accompanied by a sickening rending noise, as if every nail and
joist in the building were being tested to its limit. For the first few seconds, they watched their bedroom wall twisting from one crazy paral-
lelogram into another. Soon, however, there was a series of explosions,
accompanied by brilliant flashes of light, and the room was plunged
into complete darkness. They could do nothing but cling to each other
and hope that the earthquake would stop before the building collapsed.
One particularly severe jolt tore a 12-foot-long bookcase out of the liv-
ing-room wall and threw it and its contents across the floor.

Perhaps half a minute after its onset, the shaking did stop. Simon
and Mike felt around without success for their flashlight. Cautiously
they groped their way out of their apartment, down the glass-strewn
stairway, and out onto the cracked tiles of the patio, where their neighbors were beginning to assemble. The smell of gas was in the air—from
a house, two doors away, that had slid right off its foundations—so can-
dles were not an option. They huddled around flashlights and a portable
radio. It was a long, cold wait for the dawn—a wait marked by frequent
aftershocks and by more flashes as transformers exploded around the
city.

For many, the earthquake was a far worse experience. Thirty-three
people died as a direct result of the earthquake. Among them was
California Highway Patrol officer Clarence Wayne Dean. On his way to
work in the predawn darkness, Dean drove his motorcycle off the end
of a collapsed freeway overpass (see Color Plate 12). Another 16 people
died in the Northridge Meadows apartment building, whose first floor
was crushed almost flat by the weight of the overlying building. Besides
the direct fatalities, there was also an increase in the rate of fatal heart
attacks in the aftermath of the earthquake. Other indirect fatalities
included three people who died, weeks after the earthquake, in an out-
break of valley fever: dust clouds, stirred up by landslides in the Santa
Susana Mountains, had carried the causative organism, a fungus, into
residential areas. In addition, there were 130 injuries requiring hospital-

1   The ribbonlike scarp raised during the 1983 Borah Peak, Idaho, earthquake winds its way along the western flank of the Lost River Range. Hundreds of previous ruptures, over millions of years, have created this impressive mountain front. (Kerry Sieh and Simon LeVay)

2   Close-up of the 1983 scarp, taken 14 years after the earthquake. The sagebrush-covered slope between the scarp and the fence is the eroded remains of the scarp raised during a rupture on the same fault about 5000 years ago. (Kerry Sieh and Simon LeVay)

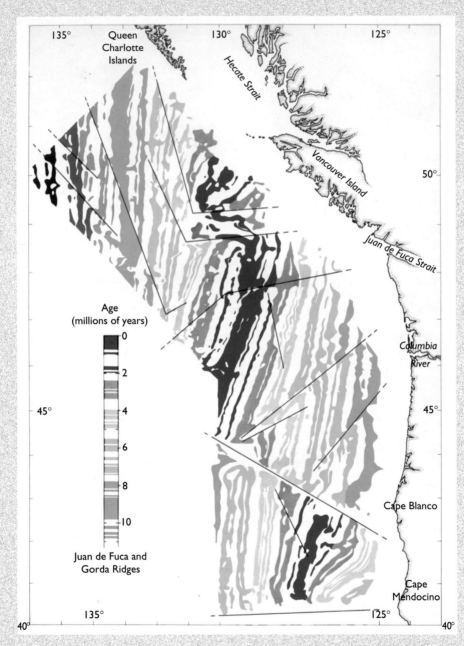

**3 Paleomagnetic stripes on the ocean floor off the Pacific Northwest.
The youngest ocean floor (red) lies along the mid-ocean ridge, and
successively older stripes of ocean floor have been displaced to the west
and east from the ridge. The black lines mark discontinuities in the
stripes.** (From an illustration by F. Vine)

4 The Turnagain Heights subdivision of Anchorage, Alaska, after the great subduction earthquake of March 27, 1964. Four minutes of intense shaking liquefied the underlying alluvium, which caused the surface to break up and slide oceanward (to the left). (Clarence Allen, Caltech)

5 Bank of the Willapa Bay estuary, Washington State, showing a buried peat layer (with embedded tree roots) that subsided during the most recent subduction earthquake. About halfway between this layer and the water is another dark peat layer that subsided during an earthquake several hundred years earlier. (Kerry Sieh and Simon LeVay)

6 Boundary between a buried peat layer (below) and a layer of silt (above), at Willapa Bay, Washington State. The three gray horizontal stripes running across the center of the photograph are bands of sand deposited during successive surges of a giant tsunami. (Kerry Sieh and Simon LeVay)

7 A row of sand-blows, formed during the 1979 Imperial Valley, California, earthquake. Ruler is 9 inches long. (Kerry Sieh and Simon LeVay)

8 Llao Rock, on the north side of Crater Lake, Oregon. The dark lava flow capping Llao Rock erupted a century or less before the great eruption that formed the caldera. Collapse of the caldera wall exposed the Llao Rock flow in cross section, including the crater from which it issued (the trough occupied by the lowest black lava).(Charles Bacon, USGS)

9 View southeastward along the San Andreas fault in the Carrizo Plain. Wallace Creek is the gully that crosses the fault, with a pronounced dogleg, near the bottom of the picture. The straight diagonal line is a fence-line. (G. Gerster, by permission)

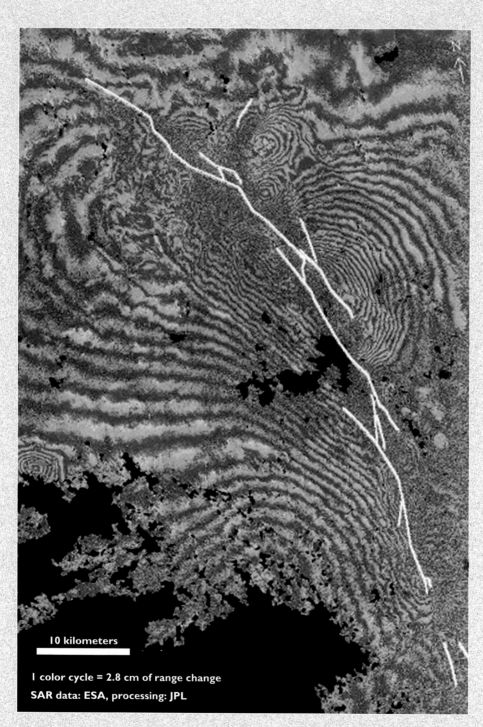

10 kilometers

1 color cycle = 2.8 cm of range change

SAR data: ESA, processing: JPL

10   Displacement of the ground surface resulting from the 1992 Landers, California, earthquake, as determined by space-based interferometry. (Courtesy of Gilles Peltzer, JPL)

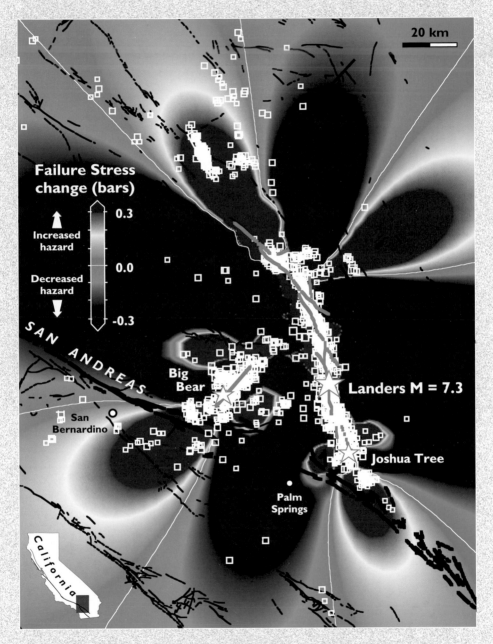

Failure Stress change (bars)

↑ Increased hazard

0.3

0.0

Decreased hazard

-0.3

SAN ANDREAS

Big Bear

Landers M = 7.3

San Bernardino

Joshua Tree

Palm Springs

California

20 km

11   Stress changes caused by the 1992 Landers earthquake rupture, calculated by R. Stein and others. The faults that ruptured during the earthquake are shown in green; others are shown in black. Regions where faults were brought closer to failure are shown in red; those where fault failure was postponed are shown in purple. Support for these calculations comes from the observation that most of the aftershocks (white squares) fall in the red zones. Stars mark the epicenters of the three major shocks.

12   Freeway bridge connecting the Golden State and Antelope Valley freeways, which collapsed in the 1994 Northridge earthquake (Jim Dewey, USGS)

ization in Los Angeles County, and many others were treated for minor injuries such as cuts. More than 20,000 people had to abandon their homes either temporarily or for good. The total economic losses have been estimated at more than $20 billion.

Kerry Sieh's experience, of course, was professional as well as personal. He and his partner were asleep at their home in Lake Arrowhead, in the mountains 90 miles east of Northridge. Although they were both shaken awake, the earthquake was a much less traumatic experience than it was for Simon LeVay, primarily because they were several times farther from the epicenter.

Radio reports soon indicated to Sieh, however, how severe and widespread the damage had been. He called Kate Hutton, whose office is just across from his own at Caltech's Seismological Laboratory, to ask the location of the epicenter—the middle of the San Fernando Valley, she told him. After a quick call to Pete Gilles, a helicopter pilot he had come to know in the aftermath of the 1992 Landers earthquake, and a cautious drive down the dark mountain to meet him at his helipad, Sieh was soon in the air headed for Caltech. And after a slightly illegal landing there to pick up a colleague and some equipment, they headed for the epicenter.

It was 9 A.M., three and a half hours after the earthquake, when they reached the San Fernando Valley. Destruction was everywhere. Buildings had collapsed, trains had derailed, fires were burning, and burst mains were spraying fountains of water skyward. At one location, in Granada Hills, Sieh photographed a scene where several fire-gutted homes surrounded a roiling fireball in the middle of the street (see Color Plate 13). The fire was being fed by a broken gas main. Incongruously, the flames were emerging through a flood of water that was pouring out of a ruptured water main. The escaping gas had been ignited by a motorist attempting to restart his engine, which had stalled as he drove through the torrent. The truck was thrown into the air and charred by the explosion, but neither the man nor his dog were hurt. It was more than a year later that USGS geologists determined that the gas and water main ruptures had been caused by liquefaction-induced ground cracking.

The main purpose of Sieh's reconnaissance was to look for evidence of ground rupture or deformation, with the hope of identifying the fault that had given rise to the earthquake. Along the Northridge Hills fault, nothing was amiss. Nor was there any sign that the fault that broke in the city of Sylmar in 1971 had moved again. In fact, aside from the ground cracking in Granada Hills, there was little evidence of a surface rupture anywhere in the valley. So, around 11 A.M., Sieh returned the helicopter and headed back to Caltech, where he joined a frenzied scientific and public-relations circus that went on almost nonstop for two weeks.

A City in a Vise

The Northridge earthquake was a reminder that the San Andreas fault is not the only seismic threat to Los Angeles. Indeed, it would be hard to imagine a tectonically less suitable place for a metropolis than the land between the San Andreas fault and the coast of southern California, for this whole region is riddled with active faults. For millions of years, and still today, rock has been slipping against rock, raising up the mountains that, when the air is clear enough to see them, give Los Angeles its singular beauty. And the mountains, eroding down not quite as fast as they are raised up, have spread themselves as gravel and sand over broad valleys. These valleys—basins filled with sediment from the mountains and from seas that came and went millions of years ago—were once home to mastodons, giant sloths, and dire wolves, then successively to scattered Indian tribes, Mexican ranchers, and Yankee fruit farmers. But now they are quilted with multiethnic suburbs, with their houses, streets, strip malls, hospitals, schools, reservoirs, and all the other fragile works of civilization. A veneer of concrete and asphalt masks the faults, warps, scarps, and all the other traces of earthquakes past, putting them out of sight and, usually, out of mind too. But rarely, a rupture tears through the veneer, changing the landscape by a geologically minuscule but still earthshaking increment.

The engine behind all this seismic activity is the same one that produces earthquakes on the San Andreas fault: the motions of the Pacific and North American plates. As we described in the last three chapters, most of the relative motion between the plates is taken up as strike-slip motion along the San Andreas and its sister faults. If these faults ran exactly parallel to the direction of motion between the plates, they would write the entire seismic story. From the creeping segment northward, the San Andreas fault does indeed run within a few degrees of the direction of plate motion, but south of the region of the Carrizo Plain, it takes a counterclockwise bend of about 30 degrees, so that the entire Mojave segment of the fault lies at that angle to the direction of plate motion. Thus, in this region, the two plates are not merely sliding past each other but also, to a lesser extent, pushing against each other. As a consequence, the entire Los Angeles area, as well as the seafloor off the coast, is gradually being compressed in a north-south direction. This compression can be measured as an ongoing process, using the GPS technique. For example, a GPS station on the Palos Verdes Peninsula (to the south of Los Angeles) is approaching the GPS station at Caltech's Jet Propulsion Laboratory in La Cañada, in the foothills of the San Gabriel Mountains, at a rate of about 8 millimeters (0.3 inch) per year. This sort of motion leads to faults of a different kind from the San Andreas fault.

The characteristic type of fault in the Los Angeles area is the *reverse fault*—a sloping dip-slip fault in which the rock on the upper side of the fault slides upward during each earthquake (see the diagram on page 66). This motion relieves the stress of compression by changing the shape of the terrain, allowing it to become shorter but higher. Typically, repeated earthquakes on reverse faults raise the land on the upper side of the fault into mountain ranges, while depressing the land on the other side into deep valleys. The net effect of a reverse-fault rupture is to thicken the crust.

A typical result of this kind of seismic activity is the range that forms Los Angeles' backdrop, the San Gabriel Mountains (see Color Plate 14). A set of reverse faults, the Sierra Madre–Cucamonga fault system, runs along the foothills of this range, from San Fernando at the western end to the Cajon Pass at the eastern end, where it meets the San Andreas fault. From its surface trace, the fault slopes down and northward, under the mountains it has raised up, and eventually meets the vertical San Andreas fault somewhere deep below the high desert. Thus, the San Gabriel Mountains form an enormous, wedge-shaped block of rock that is being popped out of the Earth's crust by the crushing pressure of the converging plates. The most recent damaging earthquake on this system was the magnitude 6.7 San Fernando earthquake of 1971, which raised a 6-mile-long portion of the mountains by 3 to 6 feet.

A kind of reverse fault that poses special hazards for Los Angeles is the *blind thrust fault*. Thrust faults are tilted so much that they are actually closer to horizontal than to vertical. Blind thrust faults do not reach the surface at all. Because there is no fault trace to be seen, the existence of a blind thrust fault has to be deduced from subtler distortions of the land. Typically, slippage on a blind thrust fault causes overlying soil and rock to tilt into a ramplike slope known as a *monocline* or to become warped around the upper end of the fault, producing a ridge known as an *anticline*. These features are recognizable even in the absence of an earthquake, but until the 1980s they were generally neglected by seismologists in southern California, simply because there had been no significant earthquakes on the underlying faults.

The Northridge earthquake was caused by slippage on a previously unknown blind thrust fault under the San Fernando Valley. Two principal lines of analysis led to its discovery: measurement of ground deformation with the GPS technique and analysis of seismographic records. Already while Sieh was circling the valley in a helicopter, Andrea Donnellan and Tom Yunck of Caltech's Jet Propulsion Laboratory were transporting a GPS receiver to the top of Oat Mountain, in the Santa Susana Range north of the valley. They found that the mountain had been lifted up by 16 inches during the earthquake. Later measurements showed that Northridge itself was raised by about 8 inches, and many other sites in the northern half of the valley were raised by similar amounts. Broad uplifts of this sort are characteristic of thrust-fault ruptures.

The fault was also delineated by Egill Hauksson and his colleagues at the Seismological Laboratory, working from the seismic-wave records. The fault plane lies at an angle of about 35 degrees from horizontal, sloping down toward the south (south-dipping). The upper, northern end terminates about 5 miles below the Santa Susana Mountains. Although the fault that produced the Northridge earthquake had not been mapped previously, it is probably an extension of the Oak Ridge fault system, a well-studied set of thrust faults that extends eastward from the coast at Ventura toward the northern San Fernando Valley.

Many faults in the Los Angeles area combine dip-slip and strike-slip motion. A series of such faults, combining reverse and left-lateral motion, runs along the base of the Santa Monica Mountains and Hollywood Hills. In West Hollywood, the glamorous Sunset Strip traverses the fault zone at the foot of the Hollywood Hills. The streets on the south side of the strip—La Cienega, Kings, Sweetzer, and Olive—fall away very steeply from the mountain front because the mountains are rising faster than debris is washing out of the canyons and into the Hollywood basin.

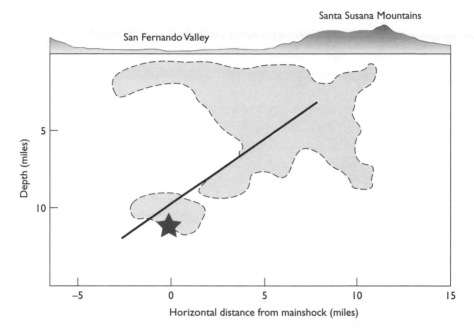

Cross section through the fault plane of the 1994 Northridge earthquake. The dashed lines surround the locations of most of the aftershocks. The star is the focus (starting point) of the rupture. South is to the left. (Adapted from USGS report)

In Hollywood, the fault runs a block or two north of Hollywood Boulevard. A good place to see the fault scarp is just above the Capitol Records building, a block north of the star-studded intersection of Hollywood and Vine (see Color Plate 15). Looking north up Vine toward the Hollywood Freeway, you can see a ramplike hill that looks as if it were constructed artificially during the building of the freeway. In reality, it is the fault scarp. If the fault were to rupture while you were standing there, you and Capitol Records would be thrown violently to the right and downward, while the buildings on the other side of the ramp would jump upward and to the left.

How is it, one might ask, that the fault has produced a scarp a mere 100 feet high or so? And how is it that the Hollywood Hills rise only a few hundred feet above the flatlands of the Hollywood basin? If the fault has been rupturing repeatedly for a long time, shouldn't there be a grander relief to show for it?

The answer is that erosion and sedimentation have largely obscured the fault's accomplishments. Almost as fast as the north side of the fault is being lifted up, the rock there is being eroded by rainfall and stream flows, and the resulting sand and gravel is being washed southward across the fault to fill the ever-deepening basin. To see what the fault would look like without these processes, you would have to take a spade to the south side of the fault and dig down through 2000 feet of alluvium and ocean sediment (for the sea once covered the land that was depressed by the fault), until you reached bedrock. Then you would have to pile all this excavated material back on the north side of the fault where it came from. You would be left with a vertiginous, overhanging cliff, several thousand feet high, from whose lip waterfalls would be cascading into the void. (Actually, those waterfalls do cascade over the fault, falling, not through air, but through alluvium to the water table hundreds of feet below.)

Assessing the Hazards and Risks

To assess the dangers posed by earthquakes in southern California accurately, one must know several things. Since earthquakes are caused by sudden ruptures of faults, we must first know the locations of the active faults. Then, for each fault, we must determine how often it ruptures and the sizes of the earthquakes produced when it does rupture. Ideally, we would also like to have some idea of how advanced the fault is in its cycle of strain accumulation and relief. Having gathered these data for every fault, we can combine them with information about how seismic waves dissipate as they travel away from the faults. Putting all this information together, we can develop an estimate of the *probabilistic seismic hazard:* the statistical probability that each location within the area will experience shaking of a given severity within a certain period of time in the future. We can then combine this estimate of hazard with knowledge about how shaking affects buildings, freeways, and other parts of our "built environment" to reach an estimate of the *seismic risks;* that is, the number of deaths and injuries, the material damage, and the economic losses that are likely to result from these hazards. Finally, these estimates of hazard and risk can provide the knowledge and the motivation necessary to reduce (or mitigate) human and economic losses through the strengthening of structures and service lifelines and through emergency planning.

The first step, locating the active faults, is reasonably well in hand in urban southern California (see Color Plate 14). It is likely that all of the larger active faults that reach the surface have already been mapped. The same cannot be said for the buried thrust faults. Still, geologists have a general sense of where these faults are. The fault that produced the Northridge earthquake, for example, had not been specifically mapped, but it lies in a region containing a number of previously mapped active faults and folds. In addition, GPS studies carried out before the earthquake had shown that Northridge and the surrounding communities lie on land that is being rapidly compressed and therefore might well contain active faults.

The next step, estimating the frequency and size of earthquakes on each fault, is more difficult, but significant progress is being made. Geologists have made excavations across several of the major faults in the Los Angeles area in hopes of revealing their recent seismic history. Tom Rockwell of San Diego State University, who has studied more than a dozen faults in southern California, is one of the most active scientists in this field. Besides helping to characterize the faults' history, Rockwell's work has produced a number of unexpected bonuses. While digging along the Rose Canyon fault in San Diego, for example, he came across a 9000-year-old Indian hearth. And while working on Santa Rosa Island in 1994, Rockwell unearthed the first complete skeleton of a pygmy mammoth—a species that became extinct (probably as a result of hunting by humans) about 11,000 years ago.

Rockwell recently collaborated with Jim Dolan and Kerry Sieh (and with geologists working for the Metrorail subway project) on a paleoseismological study of the Hollywood fault. Doing this kind of study in a city is very different from working in the Carrizo Plain or Pallett Creek, of course. Virtually every inch of the fault has been built on or covered with asphalt, and land use is so intensive that getting permission to dig a trench is well-nigh impossible. Hollywood at least offered one advantage compared with more recently developed cities: because it was constructed in the 1910s and 1920s, before the introduction of heavy earth-moving machinery, the developers "draped" the city rather lightly over the existing topography, rather than first bulldozing the entire landscape into dreary flatness. So it is possible to recognize many details of the fault's history simply by imagining the city away and inspecting the contours of the underlying ground.

For their probe of the fault zone, Sieh and his colleagues chose a quiet cul-de-sac at the end of a street named Camino Palmero, a mile and a half west of Vine Street. Camino Palmero runs up an alluvial fan (a sloping, fan-shaped surface created by the deposition of sand and gravel from a stream as it exits from a mountain range). The fault scarp is buried somewhere under the fan. Almost every location in Hollywood has some connection with movie or television history, of course, and Camino Palmero is no exception: a two-story house on the east side of the street was the locale for early episodes of the 1950s television series, "Ozzie and Harriet." Instead of cutting a trench along the street, the researchers simply drilled a series of holes in it, starting near the top of the street and moving down in intervals of about 20 feet. The drill was hollow, so that they were able to recover cores of soil and rock extending downward as much as 150 feet below the road surface.

They found that the fault runs right under Ozzie and Harriet's house. North of the house, the drill bit hit granite only a few feet down; south of it, there was no granite; the bedrock was hundreds or more feet deeper than they could drill. From the cores, they could draw certain conclusions about the date of the last rupture. A layer of sediment a few feet below the surface, which radiocarbon dating showed to be about 4000 years old, was not broken by any rupture, so the last rupture took place before then. A layer about 20 feet down, which was deposited about 20,000 years ago, *was* disrupted. So the last rupture took place between about 20,000 and 4000 years ago. Not a very precise date, of course, but enough to make it clear that, compared with the San Andreas fault, the Hollywood fault breaks very infrequently. The next rupture may well not take place until long after Ozzie and Harriet's house has crumbled to termite dust.

Although Sieh's group couldn't determine from their excavations how far the fault had slipped during the last rupture, they could make an educated guess based on the length of the fault, which is about 9 miles. For a fault of this length, a reasonable slip would be about 3 feet; this amount of slip would cause an earthquake of about magnitude 6.6—similar to the Northridge earthquake.

There is a complicating factor, however. The estimate of a 3-foot slip is based on the assumption that the Hollywood fault will break by itself. In a more alarming scenario, it might break as part of a longer rupture that included other fault segments extending all the way from the coast at Santa Monica to East Los Angeles. If the latter scenario is more realistic—and there is some reason to suspect that it may be—we could see as much as 10 feet of slip. This could produce an earthquake with a magnitude of around 7.5. The consequences of such a large

earthquake squarely within the metropolitan region hardly bear imagining, but one could reasonably expect casualties and losses 10 or more
times greater than those suffered in the Northridge event. The reasons
for the greater losses would be threefold: the greater power of the earthquake, the fact that it would affect a much larger area, and the denser
development of that area, compared with the San Fernando Valley.

The seismic history of some of southern California's faults is known
better than that of the Hollywood fault; other faults' history is even less
well known. Working from this rather spotty database, as well as from
data about the overall rate at which strain is accumulating in the region,
the Working Group on California Earthquake Probabilities has taken
on the responsibility of assessing the overall seismic hazards to the area.
The group is made up of about 15 scientists, including Kerry Sieh. In
their latest analysis, published in 1995, they calculated that the probability of a large earthquake (of magnitude 7.0 or greater) in southern
California during the next 30 years is 80 to 90 percent; in other words,
there is much better than an even chance that such an earthquake will
happen. Although it could be caused by the rupture of a fault under Los
Angeles itself, much more likely sources for an earthquake of this size
are the San Andreas fault (especially the segments near San Bernardino
and Palm Springs, which have not ruptured in the historical period) or
their neighbor, the San Jacinto fault.

The Working Group also constructed maps showing the number of
times within the next 30 years that any spot in southern California is
likely to experience an intensity of shaking sufficient to cause serious
damage. They based this calculation on the distances from each spot to
active faults, the probability and expected magnitude of earthquakes on
these faults, and the lessening of the severity of the shaking that could
be expected at those distances. The map illustrated in Color Plate 16
uses a criterion of 0.2 g, which is an acceleration one-fifth as great as
the acceleration experienced by a free-falling body. (By comparison,
some sites near the epicenter of the Northridge earthquake experienced shaking at over 1 g.)

The city of San Bernardino, lodged unenviably between two high-
probability segments of the San Andreas and San Jacinto faults, faces one
of the most serious prospects of serious shaking—a probability of over 50
percent during the next 30 years. So too does the Santa Clara Valley,
including the cities of Santa Paula, Fillmore, and Ventura. Lest others
become complacent, the map also shows that almost every populated area

in southern California, other than San Diego, has a more than 1 in 4 (25 percent) probability of being serious shaken up in the next 30 years. The map tells us that the relative seismic calm of the past century is not typical for the region; it is probably in part a result of the stress shadow of the great 1857 earthquake. The period of calm must eventually give way to a higher level of activity, and no one in southern California can consider himself or herself immune from the consequences.

Shaky Ground

In constructing maps such as the one shown in Color Plate 16, the scientists relied primarily on one kind of measure: the distance from each location to the various active faults. Long experience of earthquakes, as well as common sense, tells us that the severity of shaking lessens as the seismic waves travel farther and farther away from the rupture. But surprisingly, two sites the same distance from the rupture can experience very different levels of shaking. Sometimes, after an earthquake, a cluster of severely damaged buildings may be found in the middle of a whole subdivision of intact structures. Close study of the effects of recent California earthquakes has made clear that there are a number of factors, besides distance and the magnitude of the earthquake, that affect how severely a particular place will be shaken.

First, seismic energy doesn't radiate away equally in all directions from the fault, because the rupture is not a small focus of energy like a light bulb but an event that propagates across a large surface over an appreciable period of time. Tom Heaton of Caltech (whose work on the Cascadia subduction zone was described in Chapter 1), along with David Wald and others, has developed computational techniques to reconstruct the sequence of events during the rupture of a fault by working backward from the seismographic records and from other data. (The procedure is analogous to computing a series of CAT scans from the raw X-ray data.) What the researchers find is that the rupture moves across the fault plane as a brief pulse of fracture and rehealing. In the case of the Northridge earthquake, for example, the rupture began at a depth of about 11 miles, at the southeast corner of the fault plane, and progressed upward and toward the northwest, ending 8 seconds later at a depth of about 3 or 4 miles. At any one point, however, the two surfaces slipped against each other for no more than about 2 seconds. (You may recall from the Introduction that Mrs. Lawana Knox witnessed just such a moving pulse of rupture during the Borah Peak earthquake.)

The amplitude or size of the seismic waves radiated from the fault plane is greater in the direction of the moving rupture pulse than in other directions. This is true because the seismic waves coming from the point where the rupture begins have to travel farther to reach an observer standing ahead of the moving pulse than do the waves coming from the end point of the rupture. Therefore, the waves "bunch up" in time and space, and the total seismic energy is delivered over a relatively short period of time. In the opposite direction, on the other hand, the waves from the starting point and end point of the rupture are spread out in time, so that their amplitude is decreased.

In the case of the Northridge earthquake, the rupture pulse propagated upward and to the north. Correspondingly, the severity of shaking (as judged by accelerometer recordings) was much greater in the mountains to the north of the valley than it was in the more heavily built-up areas to the south. In fact, about the only large structure in the region of most intense shaking was the interchange between the Golden State and Antelope Valley freeways. This was the interchange whose collapse lead to the death of Officer Dean. If the fault had sloped the other way, rising to the south, the high-rises of the Los Angeles metropolitan area would have been put to a much more severe test than they actually endured.

A second reason that shaking intensity varies at different locations is that seismic waves don't always travel in straight lines. Like other kinds of waves, they can be reflected or refracted. In the 1989 Loma Prieta earthquake, both these effects conspired to increase the damage in San Francisco and the cities of the East Bay. As the seismic waves traveled the 60 miles from the rupture site near Santa Cruz to San Francisco, some waves descended deep into the Earth and were reflected back from the boundary between the crust and the underlying mantle. They reached the surface again in the vicinity of San Francisco, adding their energy to the waves that had taken a more direct path. Refraction—the change of direction experienced by a wave as it enters a medium of different density—also had dire effects: waves that started out heading for Nevada and points east slewed around as they encountered the low-density sediments of the South Bay and ended up converging on the East Bay cities. Some communities much closer to the epicenter, such as San Jose, were spared this double or triple dose of seismic energy and suffered proportionately less damage.

Reflection and refraction can cause bizarre effects when seismic waves enter enclosed spaces such as basins and hills. Waves crisscrossing within these spaces can reinforce or weaken one another, setting up large-amplitude standing waves rather like those produced in the kitchen sink when the garbage disposal is running. In the case of the Northridge earthquake, a remarkable refractive effect was observed within the deep sedimentary basin underlying Hollywood and Santa Monica. Seismologist Paul Davis and his colleagues at UCLA found that the bowl-shaped floor of the basin acted as a huge geological lens, bending the seismic waves as they entered the low-density sediment and focusing them onto certain parts of the land surface, including the city of Santa Monica. As a result, Santa Monica suffered more damage than many communities closer to the epicenter.

Seemingly insignificant wrinkles in the topography can have dramatic effects on the severity of shaking. During the Northridge earthquake the Cedar Hill Nursery, which is situated on a mere 50-foot hill in Tarzana, about 4 miles south of the epicenter, experienced horizontal acceleration that peaked briefly at 1.78 g. (Accelerating from a standing start at this rate, one would travel nearly the length of a football field in the first second. In an earthquake, however, such a high acceleration lasts much less than a second.) At the Encino Reservoir, just a mile and a half away, the peak acceleration was only one-seventh of that value. Somehow, the little hill had precisely the right dimensions to amplify the shaking sevenfold.

Perhaps the most fateful reason for local differences in shaking severity is the nature of the soil. Other things being equal, loose, uncompacted soil or landfill will always shake more than solid rock. This fact was recognized after the 1906 San Francisco earthquake: buildings constructed on landfill near the Bay suffered the most damage. The same thing happened in the Loma Prieta earthquake; many houses subsided or collapsed in the Marina District, which is built on landfill. The portion of Interstate 880 that collapsed in Oakland, killing many motorists, was also constructed on landfill. One of the most tragic examples of the effect of local soil conditions was the Mexico City earthquake of 1985. The earthquake, whose source was 200 miles away in the offshore subduction zone, killed more than 8000 people in the city, yet almost all of the deaths occurred in those small parts of the city that were built on soft lakebed sediments.

The Northridge earthquake illustrated the same phenomenon. The bridges that carried Interstate 10 over La Cienega and Robertson boulevards in Culver City, though far from the epicenter of the earthquake, were build on weak, waterlogged ground that once had been the bed of the Los Angeles River (*cienega* is Spanish for "swamp.") Both of these bridges collapsed, closing America's busiest freeway for five months.

Another factor that greatly affects the potential for damage is the possibility that the ground itself may fail, subside, or break up. Obviously, this will happen along the trace of a surface rupture. Structures, such as Ozzie and Harriet's house, that sit directly astride a fault risk being torn apart when the ground beneath them jumps in two different directions simultaneously. Blind thrust faults do not reach to the surface, but when they rupture, the surface folds that mark their presence become more sharply folded. Some of the high-rises in downtown Los Angeles sit directly atop such folds. If these faults should break, the buildings might be tilted by a degree or more, risking collapse.

Ground failure can also occur at a distance from the actual fault rupture. Most commonly, this happens on hillsides—the shaking leads to failure of the sloping ground in the form of landslides or rockfalls. During the Northridge earthquake, an expensive home located on a coastal bluff in Pacific Palisades was torn in half when the bluff failed; the seaward half of the house ended up as a pile of debris at the bottom of the slope. In the Santa Susana Mountains, where the worst shaking took place, thousands of rockslides and landslides stripped vegetation, soil, and bedrock from most of the steeper slopes.

Even rather modest ground failure can have drastic effects. During the Northridge earthquake, cracks developed along a stretch of Balboa Boulevard in Granada Hills, as the soft alluvial soil moved a few inches down the gently sloping terrain. This movement was enough to rupture two gas lines under the street, leading to the fire that was observed by Kerry Sieh from his helicopter.

For all these reasons, maps of expected shaking severity, such as the ones produced by the Working Group for the Los Angeles area, can be only a rough guide for planners. Local topography, slope, type of rock or soil, height of water table, and other factors all need to be taken into account to assess how severely any particular spot is likely to be shaken. California's Division of Mines and Geology has begun to produce maps of these landslide and liquefaction hazards in an effort to aid communities in long-range planning.

"Earthquakes Don't Kill People—Buildings Do"

Almost everyone's impulse, when an earthquake strikes, is to run outside. The Great Outdoors—meaning well away from buildings, bridges,
power lines, and the like—is indeed one of the safest places to be during an earthquake (although getting there in a hurry can be hazardous).
Landslides and tsunamis may still smother or drown you, but man-made
objects will crush, burn, or electrocute you.

Different kinds of buildings respond to shaking very differently.
One of the best kinds of building in which to safely ride out an earthquake is a modern, wood-frame, single-family home—the kind of
building in which millions of Californians live. Buildings of this kind
have the right combination of flexibility and strength to endure intense
shaking without coming apart; they glide through the oncoming seismic waves like a snake weaving through a pile of sticks. Fire, of course,
still remains a hazard, as the disaster on Balboa Boulevard illustrated.

Older (pre-1935) wood-frame homes have also done quite well in
earthquakes, although they often lack features specifically designed to
withstand shaking. Most notably, because many pre-1935 wood-frame
houses are not bolted to their foundations, an earthquake can shake
them off their foundations. This happened to many such houses in the
Northridge earthquake. Retrofitting with foundation bolts is an effective remedy for this problem, but it costs several thousand dollars and
has not generally been made mandatory.

Typically, buildings made of unreinforced masonry—adobe, brick,
stone, and the like—are too brittle to withstand intense shaking. They
rely primarily on gravity (the Earth's own g force) to hold the building
together, but during an earthquake g forces are going every which way.
(The story of the long-drawn-out struggle to strengthen Los Angeles's
masonry buildings against earthquakes is recounted in Chapter 13.)

Many people in Los Angeles live in wood-frame apartment buildings. These buildings, which are typically two or three stories high, fared
only moderately well in the Northridge earthquake. Because of the
city's reliance on the automobile, many such buildings were constructed with parking garages on ground-floor levels or lower. Parking
garages require wide entryways; sometimes one entire side of the building at the ground-floor level is no more than a series of widely spaced
posts. This kind of "soft story" construction greatly reduces a building's
ability to withstand shearing motions and therefore presents a substan-

tial seismic risk. In fact, modern apartment buildings suffered more damage in the Northridge earthquake than did masonry buildings or single-family wood-frame dwellings of any age.

Perhaps the most surprising effect of the Northridge earthquake on buildings was the extensive damage experienced by steel-frame and reinforced-concrete buildings. These buildings—the mainstay of commercial real estate in America—were thought to be well able to ride out large earthquakes because of their flexible but strong steel components. It turned out, however, that many steel-frame buildings suffered fractures in the welds that joined the horizontal and vertical beams, necessitating expensive repairs. In many instances the fractures extended into and through the adjoining beams. If left unrepaired, such damage could allow the buildings to collapse in subsequent earthquakes. (Lawyers for the company that manufactured the most widely used welding material, who are defending the company in an earthquake-related lawsuit, say that the cracks may have begun in the beams and spread secondarily into the weld material.)

One reinforced-concrete building, the seven-story Holiday Inn in Van Nuys, was severely damaged when the reinforced-concrete columns failed at the fourth-floor level. Another casualty was the reinforced-concrete parking garage at California State University, Northridge, which collapsed inward on itself—luckily, when no one was inside. The fact that failures occurred on this scale, even though (as discussed earlier) the main energy of the earthquake was directed away from built-up areas, has necessitated considerable upgrading of commercial building codes.

Although flexibility helps buildings to withstand earthquakes, it can also spell disaster. Very large, flexible buildings, such as office high-rises, respond to low-frequency ground movement with swaying motions that run up and down the building. According to one computer-based study by John Hall, a professor of civil engineering, and his Caltech colleagues, the magnitude 7.0 earthquake that geologists propose might occur under downtown Los Angeles could induce a 20-story office building to sway so violently that its vertical steel columns would break, leading to the building's collapse. So far, nothing of this kind has happened anywhere in the world. But that may be simply because a magnitude 7.0 earthquake has yet to strike a modern metropolitan area.

Because a priority is placed on preventing deaths and injuries, building codes generally emphasize the prevention of major structural collapse. But a building that stays upright after an earthquake may still have been rendered useless by severe structural or nonstructural damage. If the building serves some critical function, this loss of function may be a disaster in itself. The new Olive View Hospital in San Fernando provides a good example of this. The hospital was built in 1976, to replace a then-new building that collapsed in the 1971 San Fernando earthquake, just weeks before its official opening. In the Northridge earthquake, the new building experienced shaking that peaked at 0.91 g—far more intense than what the building was designed to withstand. Structurally, the building passed this severe test with flying colors. But the breakage of water mains within the building forced the hospital to close, right at the time when its services were needed most. Designing a building not merely to survive a large earthquake, but to do so in a fully operational condition, requires planning down to the level of every light fixture, every pipe, every shelf, and every computer terminal.

The Debate over Zoning

The current seismic zoning regulations in California were set up in 1972, following the San Fernando earthquake. The regulations focus on the location of recognizable, active faults: generally speaking, buildings may not be constructed within 50 feet of such a fault. Beyond that distance, few distinctions are made: the entirety of Los Angeles, as well as the San Francisco Bay Area, is lumped together as Seismic Risk Zone 4 (the highest risk zone) in the Uniform Building Code, which imposes building requirements such as foundation bolting, shear-resistant walls, and so forth.

Almost everyone agrees that the zoning laws need to be changed. The laws focus on a small number of very evident surface faults but disregard buried thrust faults and faults whose traces may be obscure or whose level of activity is poorly known. The laws also focus on new building construction and impose very few requirements on existing buildings, even though these buildings are likely to be around for quite a number of earthquakes.

Obviously, one cannot impose draconian measures on the basis of uncertain geology. No one, for example, is saying that Johnny Depp's Viper Room, the Coconut Teaszer, and a hundred other cultural icons along the Sunset Strip should be torn down because of a fault that may not have ruptured since Ramses II built his pyramid. Yet the Metropolitan Transit Authority, at least, takes this fault very seriously. In constructing a subway tunnel through the Hollywood Hills, the MTA engineers had to bore straight through the fault. At that location, they widened the tunnel to twice its normal width and suspended the tracks on a gravel bed for 300 feet. Their expectation is that, in the event of slippage of the fault, the tracks can be quickly reconnected, with only a gentle "kink" in the line.

The problem in Los Angeles is that the overall deformation of the basin is taken up by slippage on a large number of faults, some still undiscovered, each of which breaks very infrequently. Is the solution to this problem to continue a detailed analysis of individual faults, with the hope of defining the hazards more precisely, or should one take a global approach, tightening up building codes in a blanket fashion over the entire area?

It's a problem that has geological, economic, and psychological aspects, some of which we'll discuss more fully in Chapter 13. One reason for continuing to map and analyze the faults is that, overall, buildings in the immediate neighborhood of faults do suffer far more severe shaking than those at more distant locations, and actual ground rupture along the fault trace can have catastrophic consequences. Another reason is that, as geologists complete the inventory of faults and pin down the dates of their last ruptures, it may be possible to refine computer models of the entire fault system, such as the one developed by Steve Ward (see Chapter 5), to the point that forecasts of when and where the next earthquake will take place may be credible. In this view, we should work toward a more site-specific and hazard-specific zoning policy, with the hope of saving lives and money at the same time.

There is certainly some merit to the opposing point of view, however. David Jackson, a UCLA geophysicist and the Science Director of the Southern California Earthquake Center, believes that there are far more faults under Los Angeles than have yet been discovered. Virtually every upcoming earthquake, in Jackson's view, will be like the Northridge earthquake; that is, it will take place on a previously unmapped fault. So planning must be global, not specific: every structure should be planned with the assumption that a fault will rupture in its vicinity. Distinctions should only be made where peculiarities of the site, such as a susceptibility to ground liquefaction, affect the hazard from earthquakes regardless

of their source. The main advantage of this blanket approach is that planners are less likely to make a mistake, since they don't have to make
potentially erroneous distinctions between locations. The main disadvan-
tage is that it may be difficult to greatly strengthen building codes across
the board without encountering serious economic obstacles.

Other Cities

Los Angeles is probably unique in America in terms of the number and
complexity of active faults within its boundaries. But there are plenty
of other cities that have at least one or two such faults. San Diego, for
example, has the Rose Canyon fault, a strike-slip fault that runs north
from San Diego Bay, through Mission Bay, La Jolla, and parallel to the
shore of North San Diego County. Along the way it comes close to
many vulnerable structures, such as the numerous interchanges on
Interstate 5, as well as to the campus of the University of California, San
Diego. According to Tom Rockwell and his former student Scott
Lindvall, this fault slips at an average rate of about 1.5 millimeters or
0.06 inch per year (about one-twentieth the rate on the San Andreas
fault) and has ruptured at least twice in the past 8000 years.

The San Francisco Bay Area has a number of active faults, most of
them right-lateral strike-slip faults, each of which takes up a portion of
the sliding motion between the North American and Pacific plates.
Because the Bay Area is smaller than southern California and contains
a shorter length of the plate margin, the overall probability of a large
(over magnitude 7.0) earthquake in the next 30 years is lower—about
67 percent, according to the Working Group. Among the faults that are
of major concern are the San Andreas fault, which skirts San
Francisco's western margin, and the Hayward fault, which runs
through Oakland, Berkeley, and other cities of the East Bay. The latter
fault last broke in 1868. As mentioned in Chapter 5, there is reason to
believe that the Hayward fault is now emerging from the stress shadow of the 1906 San Francisco earthquake. The fault, as every Cal student knows, runs directly through the Memorial Stadium on the east
side of the Berkeley campus. Within a few millennia, the fault will slice
through the concrete O shape of the stadium and remodel it into a
Stanford S. More seriously, the fault threatens thousands of old masonry buildings in the East Bay cities, and it poses a special hazard to the

eastern half of the San Francisco Bay Bridge. Current plans call for this part of the bridge, which already lost one segment of deck in the Loma Prieta earthquake, to be completely rebuilt to resist an earthquake on the Hayward fault.

Seattle is another city that lies atop an active fault. As if the city didn't have enough to fear from the offshore subduction zone, a major reverse fault named the Seattle fault runs across Lake Washington, through the heart of the city, and out across Puget Sound. Robert Bucknam, Eileen Hemphill-Haley, and Estella Leopold, scientists with the USGS and the University of Washington, have found evidence that the shorelines on both sides of Puget Sound, south of the fault, were abruptly raised by as much as 23 feet about 1000 years ago. The fault rupture was accompanied by massive landslides into Lake Washington and tsunamis in Puget Sound. Bucknam and his colleagues estimate that the Seattle fault is capable of delivering a magnitude 7.0 earthquake. Because of the proximity of the fault to the city center, an earthquake of this size may present an even greater threat to Seattle's high-rises than does a magnitude 9.0 earthquake generated by the Cascadia subduction zone. Researchers have yet to determine how often this urban fault has ruptured in the past.

Besides the cities of the western United States, many other large cities around the world are exposed to the risk of earthquakes of magnitude 7.0 or greater, because of their immediate proximity to major active faults. An incomplete list would include Tokyo, Osaka, Nagoya, Kunming, Chengdu, Xian, Beijing, Bandung, Rangoon, Mandalay, Wellington, Lima, Valparaiso, Medellin, Bogota, Caracas, Managua, Guatemala City, Algiers, Athens, Bucharest, Istanbul, Damascus, Jerusalem, Tehran, Karachi, Islamabad, and Kabul.

So far in this book, we've devoted all our attention to the western margin of the North American Plate. Readers might be forgiven for thinking that this is the sole geologically active region of the country. But, important though the plate margin is, earthquakes and volcanic eruptions also occur far to the east. Our first step eastward is just across the state of California, to the ski resort that is a prime candidate to be the site of the next volcanic eruption in the continental United States.

7

THE LITTLE VOLCANO
THAT COULDN'T: FEAR
AND TREMBLING AT
MAMMOTH LAKES

"This solemn, silent, sailless sea—this lonely tenant of the loneliest spot on Earth—is little graced with the picturesque." Thus did Mark Twain describe Mono Lake, an expanse of briny, alkaline water on the east side of California's Sierra Nevada, just across the Tioga Pass from Yosemite National Park.

Twain spent a week by the lake in the 1860s. One day, as he recounts in *Roughing It,* he and a companion rowed out to Paoha Island,

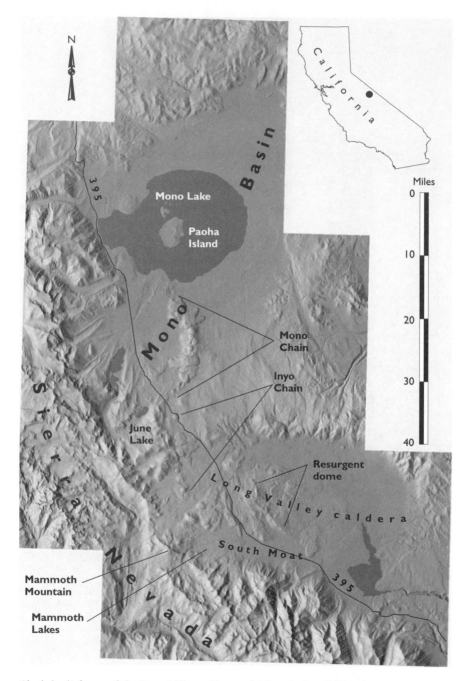

Shaded-relief map of the Long Valley caldera and Mono basin, California.

the larger of two islands on the lake. It was a hot, windless day, and by the time the men reached the island they were eager to find the fresh-water spring they had been told was located there. But there was no spring, only dusty white talc strata and a black pile of cinders, rising to a small crater from which steam jetted forth. And as the two men scoured the island for water, their boat drifted off on a rising breeze, leaving them marooned.

Our readers don't need to be told that Twain somehow survived to write *Tom Sawyer* and *Huckleberry Finn,* so we'll leave him to his predicament and take a look at the wider landscape. The eruption of Paoha, whose faint aftereffects Twain witnessed, was just the most recent of hundreds of volcanic eruptions, large and small, that have molded the terrain in the evening shadows of the Sierras. Almost everything you can see, touch, or walk upon in that neighborhood is volcanic. There are cinder cones, lava flows, craters, explosion pits, fis-sures, ash flows, and surge deposits—pretty much the entire showcase of silicic volcanism.

Today, the largest volcanic edifice in the area is Mammoth Mountain, the 11,000-foot-high ski Mecca 20 miles south of Mono Lake. But in geological terms, Mammoth is a youngster: a mere 300,000 years old. A far older, far more dangerous volcano once ruled the region from a site immediately to the east of Mammoth Mountain. Seven hundred thousand years ago, this nameless volcano erupted in a cataclysm that would have put even Mount Mazama to shame: 150 cubic miles of magma—12 Mazamas' worth—were ejected over a period of a few days. As with Mazama, the roof of the volcano sank like a piston as the underlying magma chamber emptied, leaving a 10- by 20-mile-wide depression, the Long Valley caldera. When the smoke cleared, 600 square miles of the surrounding land, including the site of the present town of Bishop, 30 miles southeast of the volcano, lay buried beneath steaming ash-flow deposits. Within the caldera itself, the deposits—the so-called Bishop Tuff—lay 5000 feet thick.

During the 100,000 years that followed, renewed volcanic activity built a resurgent dome in the center of the caldera. This dome still exists today as a broadly uplifted region a couple of miles northeast of the resort town of Mammoth Lakes. Since then, three further eruptive episodes have taken place on the resurgent dome's flanks, roughly at 200,000-year intervals; the last was about 100,000 years ago. These three eruptions, of viscous rhyolite lava, did not extend beyond the caldera.

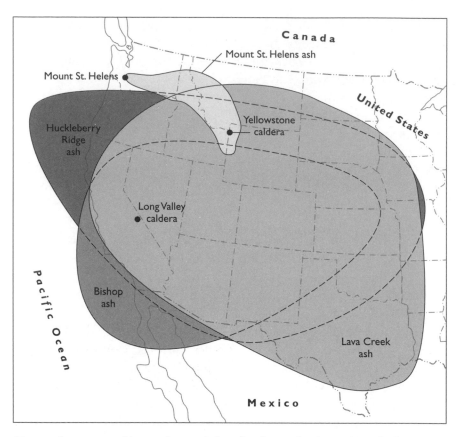

Extent of areas covered by pumice or ash from four large volcanic eruptions: the Long Valley eruption of 700,000 years ago (Bishop ash), the Yellowstone eruption of 2 million years ago (Huckleberry Ridge ash), the Yellowstone eruption of 620,000 years ago (Lava Creek ash), and the Mount St. Helens eruption of 1980. (Adapted from a diagram by R. B. Smith and L. W. Braile)

Although the Long Valley caldera has not seen a major eruption in the past 100,000 years, there has been plenty of volcanic activity elsewhere in the area, including the gradual rise of Mammoth Mountain on the caldera's western rim. Over the past 30 millennia, activity has been concentrated in a line of craters, the Mono and Inyo craters, that run northward from Mammoth Mountain to Mono Lake. Eruptions from these craters seem to happen once every several hundred years. The most recent eruption, that of Paoha Island, took place during the eighteenth century—about 100 years before Mark Twain's visit. The eruption before that took place sometime in the mid-fourteenth century. This was a double eruption: the first occurred along a 3-mile-long line of vents near the southern shore of Mono Lake; the second, which followed the first by no more than a year or so, was 11 miles farther south, near the northwestern rim of the Long Valley caldera. Before that, there was an eruption around the year 610; this eruption was along a set of vents east of June Lake, about halfway between the sites of the two fourteenth-century eruptions.

The details of all these eruptions—their dates, the kinds and quantities of materials ejected, even the wind direction on each occasion—have been established by study of the tephra beds (beds containing fragmented rocks that were thrown into the air during an eruption) and pyroclastic flow deposits they left behind. Kerry Sieh carried out one of these studies with his student Marcus Bursik; Dan Miller of the USGS conducted the other. The eruptions were dated by radiocarbon analysis of twigs and pine needles lying among the deposits or by the analysis of tree rings. To date the fourteenth-century eruption, for example, Miller and a dendrochronologist, Dave Yamaguchi, cored ancient Jeffrey pine trees growing on the deposits left by that eruption. The innermost rings of the oldest trees were laid down in 1368, so the eruption must have occurred before that. Radiocarbon dating of charred fragments in the deposits established 1325 as the earliest likely date.

A question that has not yet received a fully satisfactory answer is: Why is this part of California volcanic at all? The volcanoes of Cascadia, as we discussed in Chapter 2, have a well-defined cause—melting of the crust induced by hot volatiles rising from the subducting oceanic plate. At the level of central California, however, subduction ceased about 28–30 million years ago, at least 25 million years before the first volcano erupted in the Long Valley area.

One clue is offered by the chemical makeup of the material ejected by the volcanoes. The earliest eruptions (2.5 to 3 million years ago) were effusions of low-silica basalts that, because they were so fluid, spread in broad sheets over most of the landscape. Later, the basalts gave way to magmas that were progressively higher in silica (andesites, dacites, and rhyolites). These viscous magmas built the steep volcanic edifices such as Mammoth Mountain. The gradual progression from low- to high-silica magmas has been seen in many parts of the world and is usually interpreted as follows. The first event leading to volcanism is the infusion of hot, low-silica basaltic magma from the upper mantle into the crust.

Later there is secondary melting of the crustal rock, leading to the development of a magma chamber and to silicic eruptions.

This doesn't explain, however, why mantle material rose into the crust specifically under the region of Long Valley and the other sites of silicic volcanism around the world. There may be an element of randomness to the location of these zones. In the case of Long Valley and the Mono basin, however, a special factor may be the unusual tectonic behavior of the region. The Sierra Nevada forms the western extremity of a geological zone known as the Basin and Range province (the topic of the next chapter), which is gradually being stretched in a direction from northwest to southeast. This process has caused a thinning and perhaps a weakening of the crust; the Mono basin, for example, has sunk several thousand feet relative to the mountains to its west. This subsidence may have made the region susceptible to the infusion of mantle material.

Even if the stretching of the terrain did not directly contribute to the initiation of volcanism, the continuation of this stretching today strongly influences the form in which volcanism shows itself. Much of the magma that has reached the surface over the past few millennia has done so through sheetlike cracks called *dikes,* most of which are oriented nearly parallel to the Sierran range-front. The filling of these dikes with magma, the solidification of the magma within the dikes, and the subsequent opening of new dikes allow for the gradual expansion of the crust in the direction of the overall stretching.

The Reawakening

During the 1960s and 1970s, as the sleepy village of Mammoth Lakes grew into a destination resort of condominiums, ski shops, boutiques, and restaurants, the geology of the Long Valley caldera and the Mono-Inyo craters remained a topic of purely academic interest. But on October 4, 1978, a magnitude 5.7 earthquake, centered about halfway between Mammoth Lakes and Bishop, jolted the town.

Few people drew any connection between this earthquake and the potential for volcanic activity in the area. Nineteen months later, however, in May 1980, four magnitude 6 earthquakes occurred within the space of 48 hours, and two of these were centered within the Long Valley caldera itself. The Mammoth Lakes high school was seriously

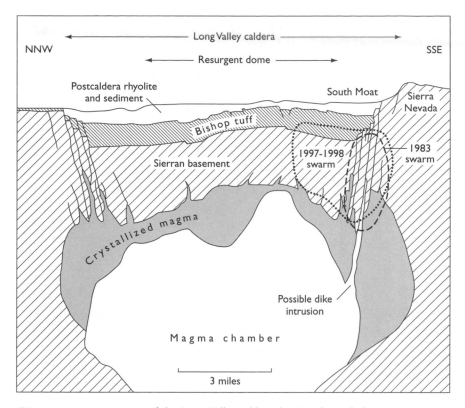

Diagrammatic cross section of the Long Valley caldera showing the underlying magma chamber, the ring fractures at its perimeter, and the locations of two major earthquake swarms. Recent findings suggest that there is not a single large magma chamber but rather a number of smaller ones. (Adapted from an illustration by David Hill, USGS)

damaged. By coincidence, Mount St. Helens had erupted a week earlier, so volcanoes were on people's minds. Could these earthquakes be precursors to an eruption near Mammoth, just as the long sequence of earthquakes at Mount St. Helens had presaged the reawakening of that volcano? This notion was soon bolstered by a subtle but ominous discovery: the floor of the Long Valley caldera had been rising. A survey conducted in mid-1980 showed that points near the old resurgent dome had risen 10 inches since the previous year.

Uplift of a caldera floor is a classic sign of the reawakening of a long-dormant volcano. It suggests that there is still a body of magma underground and that the old magma chamber is being inflated by new magma rising from the depths. In fact, the existence of a magma chamber under Long Valley had been discovered just a few years earlier. Dave Hill, of the USGS, Menlo Park, had mapped parts of it by setting off a series of explosions in the caldera floor and listening for the seismic echoes that bounced off the roof and base of the chamber. According to Hill's initial results, the chamber was a single body of magma with a volume of roughly 200 cubic miles. More recent studies, however, conducted by Hill and others, suggest that there may not be a single large body of magma after all. Rather, the gradual cooling and solidification of the magma over the ages seems to have left several smaller bodies of still-liquid magma trapped in a large mass of hardened magmatic rock (a *pluton*).

Both the earthquakes and the uplift continued. During the following two years, hundreds of earthquakes, most of them too weak to be felt, were recorded by arrays of seismographs that a USGS team had hurriedly installed. Many of the earthquakes were located below the South Moat—the region of the caldera between the central resurgent dome and the caldera's south rim. This corresponds roughly to the point where the access road from Mammoth Lakes comes down to the main range-front highway, U.S. 395. Other swarms of earthquakes were centered farther south, under the Sierra Nevada.

The South Moat earthquakes were of special interest to the seismologists, for two reasons. First, they occupied a zone of rock that is traversed by steeply sloping fractures—the remains of the ring fractures and dikes along the perimeter of the caldera, where it collapsed 700,000 years ago. Second, some of the earthquakes had seismographic waveforms that were not consistent with the usual sideways slippage of one rock face against another. Rather, it seemed that they were caused by rock being split apart, presumably by fluids (fresh magma or water) forcing their way upward from a magma chamber beneath.

The uplift of the caldera floor also continued, so that by early 1983 it was about 16 inches higher than it had been in 1979. In addition, the edges of the resurgent dome were moving apart—another sign of the expansion of a magma chamber. By 1983, geologists calculated that about 0.05 cubic mile of fresh magma had been added to the chamber.

All these observations, of course, suggested the possibility of a volcanic eruption. But the data were also puzzling. For one thing, the geological data indicated that the Mono-Inyo chain has been the site of all recent volcanism, yet the activity in 1979–1983 was centered principally around the resurgent dome of the Long Valley caldera, which hadn't seen an eruption in the past 100,000 years. Something that last happened 100,000 years ago may well happen again, of course, but it's not likely to happen next Tuesday. So most scientists would have put all their money on the Mono-Inyo chain as the site of the next eruption.

Yet the Mono-Inyo craters and the Long Valley caldera are not independent volcanic systems. The relatively young Mono-Inyo system actually extends southward into the western side of the caldera. Furthermore, magma from the Mono-Inyo system may find its way into one of the Long Valley magma chambers. Roy Bailey of the USGS's Menlo Park office has found that the lava produced by eruptions within the Long Valley caldera since the climactic eruption contains not just the long-ripening, high-silica rhyolite from the caldera's own evolutionary history, but also younger basaltic material recently injected into the roots of the magma chamber. It is possible that this material comes from the same very deep conduits that feed the Mono-Inyo volcanoes. The layout of the deep plumbing of the Long Valley and Mono-Inyo systems, and how the various volcanoes may be interconnected, is still very unclear.

Volcano Politics

On May 25, 1982, the USGS issued a Volcanic Hazard Warning for the Mammoth Lakes area. Thus began a saga that still continues: a long-running, sometimes heated conflict of interests among scientists, residents, businesspeople, and politicians over a life-threatening but reluctant volcano.

The saga got off on the worst of footings. The residents and public officials of Mammoth Lakes found out about the hazard warning not from the USGS or any other government agency but from the front page of the Los Angeles Times. The USGS had indeed mailed copies of the warning to many community organizations and leaders in Mammoth Lakes, but the Times had received a leaked copy of the memo and publicized it before the personalized mailings reached the town.

To many residents of Mammoth Lakes, the USGS had acted not merely with discourtesy but with reckless disregard for the effects the warning was bound to have on the town. And these effects were quick in coming: hotel reservations were canceled; real estate deals fell through; and restaurants, gas stations, and stores all experienced a sudden fall-off of business, right at the beginning of the town's lucrative summer tourist season.

As the weeks dragged on and the economic impact of the hazard warning became more and more obvious, many of the local businesspeople began to vent their frustration against the USGS. A community meeting was held in late August, at which time representatives of the USGS, the California Office of Emergency Services, and other agencies attempted to address the townspeople's concerns. Unfortunately, the scientists simply could not provide the information the townspeople wanted: What was going to happen, where, and when? Their caution in answering these questions seemed to take away the justification for the hazard warning. If the scientists didn't know if or when an eruption was going to happen, why had they sounded the alarm?

During the latter part of 1982, the caldera was fairly quiet, but 1983 began with a swarm of earthquakes, most of them centered in the South Moat area, nearly under the town of Mammoth Lakes and along its only access road. Two of the earthquakes had magnitudes of 5.3—enough to unnerve the townspeople, especially when they heard the scientists' analysis of the seismographic tracings. Magma, they concluded, was forcing its way as a narrow sheet or dike up the old ring fractures of the South Moat, and it had reached to within about 2 miles of the surface. At the same time, the caldera floor had risen by another 3 inches.

This time, the community reacted more positively. Mammoth Lakes was at that time an unincorporated area, so it fell to the Mono County supervisors to draw up emergency response and evacuation plans. The chairman of the board of supervisors at the time, Mike Jencks, ordered a new escape route constructed, in case the town's main access road, which ran directly above the earthquake swarm in the South Moat, was cut off. A 6-mile dirt road was ploughed through the forest to meet Route 395 well north of the old access road.

By the time of the spring thaw, the earthquake swarm had abated, and with it the townspeople's concern. Criticism of the USGS surfaced again, and in the fall Mike Jencks, as well as another supervisor who had favored a proactive response to the crisis, lost their seats in a recall election. The townspeople became focused on plans to expand the town's tourist business, and in that context volcanic hazards were better ignored. The new dirt road, which by the fall had been paved at Federal expense, had become something of an embarrassment. In a tongue-in-cheek effort to conceal its true purpose, the new escape route was dubbed the Mammoth Scenic Loop.

Another earthquake swarm hit the town in July 1984. After this swarm also subsided, seismic activity remained at a low level for five years, although the caldera floor continued to rise at a slow rate.

In May 1989, yet another series of earthquakes began, this time under Mammoth Mountain, on the caldera's west rim. A sheet of magma appeared to be forcing its way up from the depths, but it never got closer than about 5 miles from the surface. In addition, the rate at which the caldera was widening and rising increased. The activity under Mammoth Mountain declined in late 1989, but at the same time it picked up under the South Moat again. By March 1990, 100 to 300 earthquakes were being recorded every day in that part of the caldera.

Although the earthquakes under Mammoth Mountain became less frequent, there was another indication that something unusual was happening there. In 1990, Forest Service rangers noticed that trees were dying at a number of sites on and around the mountain, including the Horseshoe Lake campsite on the mountain's south flank (see Color Plate 17). At first, it was assumed that the die-off was caused by the drought that had been plaguing the area since 1986. But in March 1990, a ranger reported an unpleasant experience: as he was entering a small, snow-covered cabin by the lake, he suddenly felt as if he were being suffocated. Several similar incidents happened over the following four years, when people entered enclosed spaces such as underground utility vaults. Something toxic seemed to be in the air.

The puzzle was solved by a group of USGS scientists led by Chris Farrar; they found that the toxic agent was carbon dioxide (CO_2). This gas is normally present at low concentrations in the atmosphere. At high concentrations, however, it can kill tree roots and asphyxiate humans. Farrar and his colleagues found CO_2 levels in the soil and in confined spaces ranging from 20 to 90 percent—levels that are lethal to plants and animals alike. Within the healthy parts of the forest, in contrast, CO_2 levels were below 1 percent.

Several lines of evidence indicated that the CO_2 was coming from inside Mammoth Mountain. First, levels were highest under snow-covered ground, as if the snow were preventing the gas from escaping into the atmosphere. Second, the locations affected by the tree kills were near known fault traces, suggesting that the gas was percolating up along the fault planes from inside the mountain. But the clincher came when the scientists analyzed the gas more carefully. Several of its characteristics, such as the ratio of different isotopes of carbon in the carbon dioxide, were characteristic of gas released from magma, not of carbon dioxide derived from the air or from biological processes.

The CO_2 release means that magma is present deep within Mammoth Mountain. Either magma rose toward the surface during the 1990 earthquake swarm (which would be consistent with the seismographic data) or the earthquakes made the rock more permeable to magmatic gases that had been collecting underground over a long period.

Two features of the gas release were somewhat puzzling. First, the CO_2 was accompanied by little or no sulfur dioxide (SO_2), which is usually abundant in volcanic emissions. Second, the gas was cold. According to Hill, both of these features could be explained by assuming that the gases passed through layers of groundwater on their way to the surface; the water would cool the gas and remove the SO_2, which is highly water-soluble.

Since 1990, the release of CO_2 has continued unabated, and more and more trees have died. By itself, however, the CO_2 release says little about the prospects of an eruption in the near future. Furthermore, the most likely kind of volcanic event on Mammoth Mountain would be relatively innocuous steam blasts, produced by the interaction between magma and groundwater. Such blasts occurred in about 1350, toward the end of the most recent eruption of the Mono-Inyo chain. These blasts left explosion pits scattered at various locations around the mountain. No actual magmatic eruption has happened on the mountain for about 50,000 years.

In 1991, in an attempt to clarify its system of volcano warnings, the USGS developed a scheme in which future announcements would be accorded a letter status, from E to A, corresponding to increasing degrees of danger. Thus an E notice meant "weak unrest—or maybe it's just our instruments acting up," while an A notice meant "an eruption is likely within hours or days." The system was intended to help residents and officials better distinguish major from lesser hazards, but there were still problems. A mere level D notice, for example, issued in June 1996, provoked a report in the *Los Angeles Times* headlined "Volcano Watch Declared in Mammoth Area." Although the article itself accurately described the level of hazard, the headline, with its image of sleepless scientists glued to their seismographs, was inappropriate. In

1997, the USGS introduced a new set of color-coded notices, intended to further simplify communication with the public.

Given the fact that earthquakes and ground deformation have been going on at Mammoth for nearly 20 years without culminating in an eruption, scientists have had to think again about some of their assumptions. How strongly do seismic activity and ground uplift predict an eruption? In the case of volcanic edifices like Mount St. Helens, the answer is probably that they do predict an eruption quite strongly. But the situation with volcanoes that have formed a caldera may be different. The main problem is that episodes of seismicity and uplift may leave few traces in the geological record, so it's difficult to tell whether such episodes have occurred previously in the Long Valley caldera's past or whether this is the first such episode since the last eruption.

One way to evaluate the situation is to look at the behavior of other calderas around the world. There are several other calderas that have shown major seismic activity and uplift in recent decades. The floor of Yellowstone caldera rose very gradually, without an eruption, from 1923 until 1985, when it began subsiding again (see Chapter 9). A caldera named Campi Flegri, near Naples, Italy, started rising in the early 1970s. In a two-year period between 1982 and 1984 it rose by more than 5 feet. The uplift then slowed and stopped, without an eruption, and the floor of the caldera has recently begun to subside again. The most recently active caldera has been Rabaul, on the island of New Britain in Papua New Guinea. Uplift and seismicity at Rabaul also began in the early 1970s. During one night in 1994, the shoreline rose by 20 feet, and the next day two separate mountains, one on either side of Rabaul harbor, erupted. (Luckily, there was time to evacuate the population.) Another caldera, Newberry caldera in central Oregon, has been rising slowly for hundreds of years, without an eruption and with very little seismicity. Thus, uplift and earthquake swarms may culminate in an eruption, or they may peter out uneventfully. It may be many decades before we know which outcome is Long Valley's destiny.

Although a volcanic eruption without detectable precursory events is rare, the practical value of such warnings is lessened by the unpredictable time relationships between the precursors and the eruption. In the case of Mount St. Helens, the precursory signals culminated in an eruption only two months after the initial tremors. The story was quite different on the Caribbean island of Montserrat. On three separate occasions in 1995 and 1996, ominous signs of activity within the

Soufrière Hills volcano led to the evacuation of Plymouth, the island's capital, but these signs were not immediately followed by significant eruptions, and some residents of Plymouth returned to the town. Then, in June 1997, the volcano let loose a pyroclastic flow that killed at least 12 people and caused severe damage in Plymouth. Two months later, the volcano erupted even more violently, and the entire southern half of the island had to be evacuated.

David Hill is very conscious of this same problem at Mammoth. "The thing we worry about," he says, "is whether we will be too conservative or too radical. The message we communicate may cause people to be evacuated for months. And if they come back and nothing has happened, it's going to be very difficult to move them again, unless there are very noticeable things going on in the Earth."

There is one scenario by which an eruption at Long Valley or the Mono-Inyo chain might occur with little or no warning. This would be if an eruption were triggered by a large tectonic earthquake. There have been examples elsewhere of such earthquake-triggered eruptions: the great Chilean subduction earthquake of 1960, for example, was followed two days later by the eruption of the Andean Puyehue volcano. There are plenty of potential sources for a large earthquake near Long Valley. In 1872, the third largest earthquake in California's recorded history—an estimated magnitude 7.6—rocked Owens Valley, just 50 miles down Highway 395 from Mammoth. Even the 1992 Landers earthquake, a slightly smaller earthquake centered 280 miles away from Mammoth, was followed by a temporary increase in seismicity within the caldera. Thus, any large earthquake that happened near Mammoth would have the volcanologists anxiously watching the caldera over the following weeks.

Besides the uncertainty about if and when an eruption will occur, an even greater uncertainty surrounds the question of what form the eruption might take. It could be anything from a few steam blasts on the flanks on Mammoth Mountain to a repetition of the climactic eruption of 700,000 years ago. But the most likely scenario is something in between: a modest eruption within the caldera or along the Mono-Inyo chain.

Even a small eruption, of course, could have devastating effects on Mammoth Lakes. In drawing up emergency response plans, the USGS and the California Office of Emergency Services (OES) have had to deal with the possibility that their response teams would be forced to leave the town. In the OES's plan, its headquarters will be located at Bishop. This says something about the OES's assessment of the likelihood of a repetition of the event of 700,000 years ago: if a similar eruption were to happen again, their headquarters, along with the entire town of Bishop, would be reduced to a stratigraphic smudge between "Bishop Tuff I" and "Bishop Tuff II."

The USGS, in contrast, has sited its fallback headquarters at Bridgeport, on higher ground 50 miles to the north of Mammoth. Not that the USGS takes the likelihood of a repetition of that climactic event very seriously. But, according to Hill, even a smaller eruption in the wrong circumstances could endanger Bishop. For example, an eruption of one of the Inyo volcanoes during a period of heavy snow cover could cause rapid melting of the snow. The resulting mudflows might travel many miles downhill toward Bishop.

The city of Mammoth Lakes (the community is now incorporated) has primary responsibility for the safety of its residents in the event of an emergency. Glenn Thompson, who was city manager from 1990 to 1996, told us that the city has developed plans for a variety of possible scenarios, most of which anticipate that the caldera will give due notice of an upcoming eruption. Even in the worst case—an eruption without notice on a busy weekend, when 30,000 people are crowding the town—Thompson believes that the town could be evacuated rapidly and safely. A group of about 100 citizens has been specially trained to help the police, fire, and Forest Service officials notify the population: they would alert people in every dwelling, business, and campsite. In general, people would be expected to drive themselves out of the town, but the buses belonging to the school district and the ski area would also be available. Thompson emphasized that there's nothing unusual about a city having emergency plans to evacuate the residents—only the nature of the hazard is unique to Mammoth Lakes.

The Spin

The current situation is a kind of uneasy truce among the three main parties: the USGS scientists, the citizens of Mammoth, and the caldera. USGS scientists continue to monitor the volcanoes from their offices at Menlo Park, but they do not spend much time at Mammoth. The townspeople have generally lost the urgent interest in the matter that they had in the early 1980s; their main hope is that Mammoth's status as an active volcanic area will recede from the world's consciousness. So when the *Mammoth Times,* for April Fools' Day in 1996, ran a story announcing that an eruption was imminent, many townspeople were not amused, especially when the story was picked up by the *Los Angeles Times* and people began calling to ask if it was true. The *Mammoth Times* had to do some quick damage control. "The opinion of Dr. David Hill," it declared in the following issue, "is that any type of local volcanic activity in our lifetime is extremely unlikely."

Although real estate prices remain depressed, development is taking off again. A Canadian investment company, Intrawest, has purchased a large section of Mammoth Mountain and plans to construct a new base lodge, a golf course, and timeshare condominiums. The developers' due-diligence studies, carried out before the purchase, persuaded them that their investment was not unduly threatened by seismic or volcanic activity in the caldera.

Sam Walker, who owns a brewery and a restaurant and has been chairman of the chamber of commerce, told us that relations between the townspeople and the USGS scientists had improved greatly in recent years. A lot of the misunderstanding, he told us, was caused by inaccurate reporting of what the USGS people had said. If the USGS said that an eruption was a possibility, some TV station or other would announce that lava was flowing down the streets. "The media are struggling for ratings," he told us. "I guess they need to sensationalize everything, but when it's your livelihood and your family's livelihood at stake, you're a bit sensitive to that."

Walker's opinion was seconded by Glenn Thompson. "That reporter down there who has such a fascination with us," he said, referring to Kenneth Reich of the *Los Angeles Times,* "most of the time he does a fairly accurate job. But he makes some errors too, because he has historically felt that we're hiding things. And the television is the worst: what they do, even if it's just a minor monitoring change, is to run footage of Mount St. Helens going off in the background, so even if what they're saying is true, they distort it through the video portion."

Walker still feels that the USGS scientists unwittingly exaggerate the risks. "David Hill will tell you the likelihood of an event in the next 200 years is fairly high," he said. "But what's an 'event'? He means anything from a Mount St. Helens all the way down to some hot water coming up in a stream. And throw into that the question of where it might happen; the likelihood of it being anywhere near any populated area is fairly remote. You can feel pretty safe that nothing is going to occur that could have any significance to your property or your family. So why doesn't he use *that* likelihood? All the media hears is that something is going to happen in 200 years."

Nancy O'Kelly, who owns a restaurant and is president of the Rotary Club, also believes that the geologists have a communication problem. "Their language is much different," she told us, "in that they talk in long periods of time. 'Imminent' does not mean imminent to them. The community and a lot of news media never really understood that." And she agreed with Walker that there was confusion about the expected size of the event, should it occur. "I've talked to Dave Hill about this," she told us, "and a lot of times the eruption would be more of tourist interest— it wouldn't be as devastating as they make it out to be."

Although Hill's opinion, as filtered through the minds of Mammoth residents, is that nothing of any consequence is likely to happen any time soon, Hill's own pronouncements are far more guarded. He would like to be able to make a quantitative estimate of the probability of an eruption, as is increasingly being done in the field of earthquake forecasting. As a starting point, he points out that over the past few millennia, magmatic eruptions have taken place at intervals of about 200 to 700 years. The "background probability" of an eruption in any given year is therefore one in a few hundred. This background probability has to be raised, however, to take into account the current unrest in the caldera. How much it should be raised is the fundamental problem, which presently defies solution. Hill hopes that further study of the Long Valley caldera and other calderas around the world will allow this missing part of the equation to be filled in.

Meanwhile, the caldera continues its pattern of slow uplift and occasional earthquake swarms. A particular vigorous swarm began in the summer of 1997. By late November, earthquakes were occurring at a rate of 2300 per week, including shocks of magnitude 4.8 and 4.9. The swarm was accompanied by an increased rate of deformation of the caldera. At the time of this writing (early 1998), there is no sign that the swarm is abating. The restless magma seems to be considering its options: whether to retreat into the volcano's deep recesses, wipe Mammoth Lakes off the map, or throw up some well-tempered fire fountain—a picturesque but unthreatening spectacle that will ensure the resort's well-being for decades to come.

As mentioned earlier, Mammoth Lakes sits on the western flank of a 600-mile-wide tract of rugged terrain known as the Basin and Range province, which is currently widening at a rate of about half an inch a year. This process of widening leads to a characteristic style of geological faulting and makes the province into prime earthquake country, as we will see in the next chapter.

8

INLAND MURMURS: EARTHQUAKES IN THE BASIN AND RANGE PROVINCE

It was a close, warm, breezeless summer night. The light of a full moon played on the 15-mile expanse of Hebgen Lake, a popular resort area just west of Yellowstone National Park. Beyond the earthen dam at the lake's north end, the picturesque gorge of the Madison River was dotted with vacationers' campsites. The Bennett family had stopped there that day, on their way to Yellowstone from their home in Coeur d'Alene, Idaho. Their trailer was parked among pine trees near the

A landslide caused by the 1959 Hebgen Lake earthquake blocked the Madison River gorge, resulting in the formation of Earthquake Lake. (U.S. Forest Service photograph)

Rock Creek campground—a particularly scenic location where the river squeezes its way between 2000-foot-high rocky slopes. The Bennett parents were already asleep in the trailer, while their four children were in their bedrolls on the ground outside. It was 23 minutes before midnight on August 17, 1959.

Afterward, Mrs. Bennett didn't recall the earthquake—either she slept through it, or the horror of what followed erased it from her memory. What she does remember is being awakened by an indistinct sound, followed by a great roaring noise. When she and her husband stepped out of the trailer to see what was happening, they were struck by a powerful blast of air. Her husband was picked up by the wind; he managed to clasp onto a tree, but the wind was so strong that he was blown horizontal, like a flag, and within a few moments he lost his grip and was blown away. Mrs. Bennett also saw one of her children blown away from her, followed by a car, rolling over and over like a log. She blacked out. Some minutes later, her 16-year-old son Phillip, who had also been tumbled about by the blast, found himself in water. Although one of his legs was broken, he managed to claw his way to some trees, where he remained all night, half-buried in mud. Phillip and his mother were the only members of the Bennett family to survive that night.

The Maults, an elderly couple from California, were also asleep in their trailer near the Rock Creek campground when the earthquake struck. Suddenly, they found themselves trapped within their trailer as it was carried along on a flood of water. Strangely, the water carried them upstream, toward the dam, not away from it. When the trailer came to rest, the Maults barely had time to struggle outside before it was submerged. They climbed from the roof of the trailer into the boughs of a nearby pine tree. As the water level continued to rise during the night, they had to climb higher and higher. Several times branches broke and tumbled them into the water, but each time they made it back into the tree. At first light, they were finally rescued.

The Ost family was sleeping in a tent at the Rock Creek campground. They were awakened by the intense shaking of the ground and scrambled out of their tent to see what was happening. A few seconds later, they heard a tremendous grinding noise and the sound of rushing water. Mr. Ost saw that the water in the Madison River was hurtling *upstream* and overflowing its banks. The campground was soon inundated by the floodwaters, but the family members saved themselves from being swept away by hanging onto trees. Eventually, they were able to climb to higher ground.

What bizarre event or events had triggered these catastrophes in Madison Canyon, below the dam of Hebgen Lake? What caused the blast that blew away the Bennetts? Why were the Maults carried upstream, toward the dam? And why were the Osts inundated so rapidly?

The answer became apparent at daylight. The survivors and their rescuers realized that an enormous landslide, triggered by the earthquake, had crashed down the south slope of the gorge, several miles below the dam, and blocked the river. According to later estimates by geologists, nearly 40 million cubic yards of rock and debris slumped

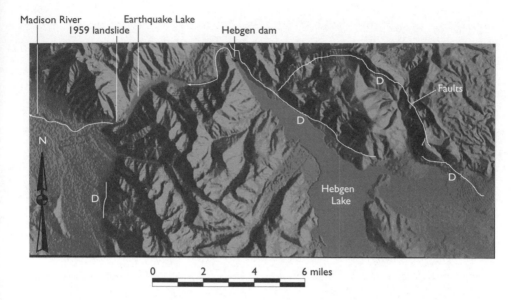

Madison River   Earthquake Lake
    1959 landslide | Hebgen dam

Shaded-relief map of Hebgen Lake area showing the features referred to in the text. The D next to each fault indicates the side that moved downward during the 1959 earthquake.

onto the river and the highway, choking the gorge with rubble nearly 200 feet deep. By suddenly displacing the air and water in the gorge, this landslide had caused the extraordinary blast of air and the river's backward flood.

With the river blocked, a large body of water began to accumulate on the upstream side of the landslide. By the morning after the slide, Rock Creek campground was deep under water. Three weeks later, "Earthquake Lake" was nearly 200 feet deep and extended for 6 miles upstream, almost to the foot of Hebgen Lake Dam. Fearful that the new lake, once it crested the craggy surface of the landslide, would erode a gorge through the slide and generate a catastrophic deluge downstream, the Army Corps of Engineers worked nonstop for days to bulldoze a wide, flat spillway across the slide. The level of Earthquake Lake stabilized at the level of the spillway, nearly 100 feet below where it would have been without human intervention.

Miles upstream from the landslide, Hebgen Lake itself was greatly disturbed by the earthquake. At the moment of the shock, the floor of most of the lake dropped by as much as 10 feet, while the lake's southern bays were raised by about 5 feet. Thus, for a few seconds, the lake's surface was tilted out of the horizontal. The water responded, of course, like the water in a bathtub that has been lifted at one end: it surged toward the lower end, was carried by its own momentum past its new equilibrium position, and raced on toward the dam. By 10 minutes after the earthquake, the water had reached the dam and overtopped it, forming a short-lived waterfall that spilled down the dam's earthen wall and into the Madison River. As fast as it had risen, the water retreated, leaving the dam crest high and dry. At intervals of several minutes, three more surges crossed the crest of the dam. The water, along with the shaking and subsidence, severely damaged the dam's earthen walls, the sluice gates, and the spillway, but the dam did not fail. The lake surges continued to crisscross the lake for at least 12 hours after the earthquake was over.

The Hebgen Lake earthquake had a magnitude of 7.3 and resulted in the deaths of 28 people. Most of the victims had been camping either at Rock Creek, just upstream from the landslide, or farther downstream, directly in the slide's path. Every building in the neighborhood of Hebgen Lake was damaged, many beyond repair; highways were knocked out by landslides or subsidence; and water pipelines were broken in many places. The earthquake was felt over an area of 600,000 square miles.

The earthquake at Hebgen Lake was caused by the rupture of several *normal faults*—a kind of fault that we have not yet described in detail (see the diagram on page 66). Like a reverse fault, a normal fault is a dip-slip fault; during a rupture, the two faces slip vertically against each other. But the direction of slip on a normal fault is the exact opposite of that on a reverse fault; the upper face (or "hanging wall") slides downwards, relative to the lower face (or "footwall"). Thus, if one stands on the hanging-wall side of a normal fault after a rupture and looks across the fault, one will see a fresh scarp, perhaps a few feet high—a strip of the footwall that was exposed to the air as the hanging wall slid downward.

Readers will now understand why "reverse" faults are so named—their slip direction is the opposite of that shown by normal faults. But what is "normal" about normal faults? The nomenclature is really just a historical accident: they were the common type of fault encountered by coal miners in eighteenth-century Britain. The miners noticed that if the horizontal vein of coal they were following came to an abrupt end at a fault, they could usually pick it up again by following a simple rule of thumb: dig downward if the fault leans toward you as it rises (that is, you're approaching it from the footwall side) and upward if the fault leans away from you (that is, you're approaching it from the hanging-wall side). This was the "normal" pattern. The rare faults that violated this rule were "reverse" faults.

Like reverse faults, normal faults often run along the base of mountain ranges that owe their existence to repeated ruptures of the fault. Beneath the surface, however, the normal fault slopes downward under the valley, not back under the mountain range as with reverse faults. After the Hebgen Lake earthquake, several normal fault scarps were found to have ruptured. Most of these ran parallel to the east side of the lake, along the base of the Madison Range. The faults sloped westward under the lake, so that the lakebed sank as the fault ruptured.

The Pioneer

Many normal faults slice up the large, arid region of the United States that lies between the Sierra Nevada, to the west, and the Rocky Mountains and Colorado Plateau, to the east. This region—known to geologists as the Basin and Range province—includes most of Nevada, the western half of Utah, and parts of the adjoining states to the north and south (Hebgen Lake is actually in the southernmost tip of Montana). The central part of the province—the Great Basin—is so rimmed in by high mountains and valleys that rivers have no exit to the ocean; instead, they meander to their termini in huge white salt wastes, such as the Humboldt and Carson sinks east of Reno, Nevada, and the Great Salt Lake Desert west of Salt Lake City. These salt pans, and the Great Salt Lake itself, are the desiccated remnants of immense bodies of fresh water that once graced the region. From 30,000 to 13,000 years ago, when ice covered most of Canada and the northern Great Plains, the Great Basin was wet and cool enough to support two giant lakes: Lake Lahontan and Lake Bonneville. Lake Bonneville was 1000 feet deep and as large as Lake Michigan is today.

The "basin" in Basin and Range refers not to the Great Basin but to the innumerable little basins—10- or 15-mile-wide valleys—that separate the region's mountain ranges into an almost regular corrugation. Because the ranges in much of the province run north and south, collectively they impose a formidable barrier to east-west travel. The

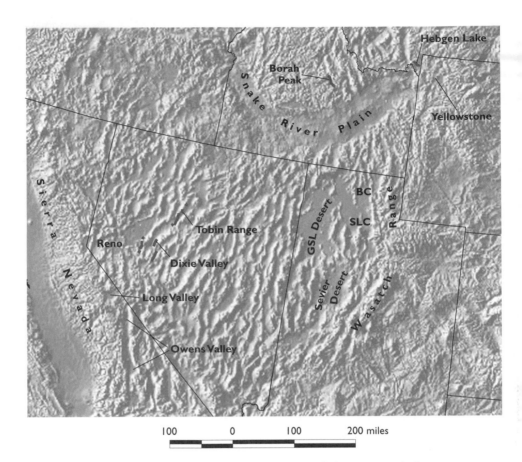

Shaded-relief map of the Basin and Range province showing fault ruptures and places mentioned in the text. SLC = Salt Lake City; GSL = Great Salt Lake; BC = Brigham City.

challenge of the terrain can be appreciated by driving (or better yet, bicycling) along one of the east-west highways, such as U.S. Route 50 (the old Pony Express highway that traverses the entire province), from the Wasatch Mountains in central Utah to the Sierra Nevada on the western edge of Nevada. For those without time or inclination for such a trip, we've included a computer-generated relief map of the entire region. This view makes clear why the Basin and Range has been likened to a "herd of caterpillars migrating northward."

Grove Karl Gilbert, generally considered the greatest of nineteenth-century American geologists, first saw the Basin Ranges from a train. He journeyed through the region in April 1871, two years after a golden

spike had been hammered into a wooden tie north of the Great Salt
Lake to mark the joining of the Central and Union Pacific segments of
the first transcontinental railroad. The 27-year-old geologist was traveling to San Francisco to join a surveying party. As range after range of
mountains rolled past his carriage window, Gilbert was struck by the
peculiar regularity of their structure. He could see that each range was
composed of tilted layers of rock. On one side of the range, the strata
followed the slope of the mountainside, forming a shallow ramp. On the
other side, the layers were abruptly broken off. For several ranges in a
row, the rock strata were tilted in the same direction, like row after row
of canted skylights on a factory roof. Then a series of ranges might follow in which the strata were tilted in the opposite direction. What, he
wondered, was causing these patterns?

In those days, American and European geologists believed that
mountains were formed of *folds*—wavelike wrinkles in the crust that
were evidence of the crust's contraction. Indeed, folds had created the
mountain ranges in the parts of the world where the geologists had
been trained, such as the Appalachians in the eastern United States and
the Alps in central Europe. And so the early geologists who studied the
Basin Ranges assumed that they too must have been formed in this way.
Some horizontal compressive force must have wrinkled the terrain, and
then one side of each wrinkle must have been eroded away, leaving a
ramp or, as geologists call it, a *monocline*.

Gilbert had a hard time imagining how erosion could have given
rise to such patterns. Over the next two years, he took the opportunity to examine many of the ranges more closely. He noticed that small
scarps ran along the base of many of the ranges on their "broken-off"
sides. Some of the scarps were clean and fresh, like those formed in
1959 at Hebgen Lake; others were more or less blunted or obliterated,
as if a great deal of time had passed since they had formed. To Gilbert,
the scarps bore witness to the most recent events in a long series of dip-
slip ruptures, which collectively enabled the layers of rock to tilt thousands of feet into the air.

The Basin and Range province, Gilbert realized, had long ago
begun to break into a set of tilted blocks, like the volumes of an ency-
clopedia when one bookend is removed. Just as the smooth top surface
of the set of books becomes ribbed like a washboard as the volumes tilt
farther and farther over, so the originally level province acquired a saw-
tooth profile as the blocks sheared past each other.

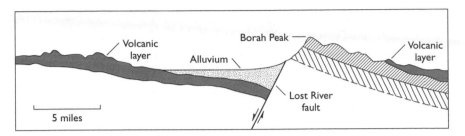

Cross section through the fault zone of the Borah Peak earthquake, showing the offset of strata on the range-front normal fault. (Adapted from an illustration by R. S. Stein and others)

If this was the correct explanation for the formation of the basins and the ranges, it should be possible, conceptually, to push the books back upright—that is, to imagine reconstructing the original level surface of the region by reversing the accumulated slip on the normal faults. Although Gilbert didn't try to do this, others have done so since. A number of geologists, for example, have studied the layering along the Lost River Range in Idaho, site of the 1983 Borah Peak earthquake witnessed by Lawana Knox (see the Introduction). Here, the original surface was a volcanic deposit—part of a huge outpouring of basaltic magma that covered most of the area about 10 million years ago, around the time that faulting began. In the range west of the Lost River fault, this volcanic layer still lies on the surface. But as one approaches the fault, the volcanic layer dips down under thick alluvial deposits, meeting the fault about 7000 feet underground. On the east side of the fault, by contrast, the volcanic layer has been raised upward to form the top of the Lost River Range. Over time, it has been completely eroded away from the top of Borah Peak and is found again only on the lower eastern slopes. Rough calculations indicate that, to get the volcanic layer into alignment across the fault, one would have to slide the two faces of the fault backward by more than 3 miles.

Because the 1983 earthquake was accompanied by about 6 feet of slip on the fault (and a similar 6-foot slip happened a few thousand years ago), one can make a rough estimate of the number of such ruptures it would have taken to raise the Lost River Range. It works out to about 2700 ruptures, each producing an earthquake of about magnitude 7. As the whole process took something like 7 million years, such ruptures would have happened at a rate of about one every 2600 years.

Of course, this is just a thought experiment to give us some notion of the immensity of time that must have been involved and the inexorable nature of the process—we don't really know that all the seismic building blocks were identical or evenly spaced in time.

Gilbert understood the large-scale implication of his "block-tilting" theory: the Basin and Range province must be getting wider. The mountainous terrain was not the result of contraction, as his contemporaries had supposed, but of its exact opposite: extension. Today, with the aid of the GPS technique, the extension of the Basin and Range province can be followed nearly as it happens—or as scientists say, in "real time." If the Rocky Mountains are taken as stationary, the Sierra Nevada is moving northwest at something like 10 millimeters (or 0.4 inch) per year. And as the province stretches, the crust must be getting thinner.

Gilbert couldn't offer any useful explanation of why this large-scale stretching process is going on; in fact, he was more focused on the local events that built the individual basins and ranges than on the evolution of the entire province. Even today, the question hasn't been fully answered, although there are a number of ideas. According to one notion, the province is stretching as part of a recovery from an earlier period of contraction and thickening—perhaps in the same way that Tibet, having been raised to giddy heights by the collision of India and Asia, is now beginning to spread itself out to the east and west. Alternatively, the current stretching may be related to the collision of the North American and Pacific plates: about a fifth of the relative motion between the two plates is taken up by the northwestward motion of the Sierra Nevada. In this conception, the plate boundary is even fuzzier than we indicated in Chapter 3. Straddling the San Andreas fault doesn't quite make you the Paul Bunyan of plate tectonics: you can't claim to have one leg securely on the North American Plate until you plunk it down on the ski slopes east of Salt Lake City.

Making the Connection

Besides explaining the formation of the Basin and Range, Gilbert's fieldwork led him to an even more fundamental discovery: that earthquakes are caused by the rupture of faults. Today, that is usually presented as so obvious a fact as barely to require documentation. But in the 1870s it was far from obvious. People knew that earthquakes were often accompanied by cracking of the ground, but they thought of the faulting as the *result* of the earthquake, not its cause: the Earth simply shook so hard that it broke. Even nowadays, we often hear seismologists saying that a particular earthquake caused the rupture of such-and-such a fault, even though it was really the other way around.

As to why the Earth shook in the first place, opinions were many and various. In early human cultures, such as that of the Klamath Indians (see Chapter 2), earthquakes and volcanic eruptions had been seen as spirits; that is, they were independent portions of the Will that suffused all Nature. In premodern Western culture, they were no longer thought of as having their own volition but rather as being the instruments of an external deity. Thus, earthquakes might be visited on humankind as retribution for some sinful behavior.

The notion that earthquakes had discoverable physical causes, although first put forward by the philosophers of ancient Greece and Rome, didn't really catch on until the eighteenth century. Even then, the matter remained very much in dispute, especially in the aftermath of the terrible earthquake that destroyed the city of Lisbon, Portugal, in 1755. Jean-Jacques Rousseau, the French philosopher, suggested that the earthquake was God's retribution on the citizens of Lisbon for choosing to dwell in seven-story houses, when He had intended them to live in a state of Nature among the trees. Rousseau's opponent, Voltaire, was skeptical. "One longs, on reading your book, to walk on all fours," he commented. "But as I have lost that habit for more than sixty years, I feel unhappily the impossibility of resuming it."

By the 1870s, the common belief among geologists was that earthquakes were caused by underground volcanic explosions. We now know that the movement of magma beneath volcanoes, such as those at Mount St. Helens and the Long Valley caldera, can indeed cause earthquakes, but the great majority of earthquakes happen in regions where there is no volcanic activity. So some other hypothesis is necessary.

Gilbert was fortunate to have visited the Owens Valley, at the western edge of the Basin and Range province, only a short time after it had been struck by the great earthquake of March 26, 1872. There he saw the fault scarps that had appeared during that earthquake, as well as the tremendous damage caused by the shaking: the town of Lone Pine, near Mount Whitney, had been leveled, with many fatalities. (One can still visit the mass grave of 23 victims, at the north end of town just off the main highway.) Gilbert also noticed that there were signs of earlier, similar ruptures along the same scarps. Putting together these observations and his broader studies of the Basin and Range, Gilbert came to the idea that the ranges were built through the slow accumulation and sudden release of strain in the rocks. "The instant of yielding," he wrote in 1873, "is so swift and so abruptly terminated as to constitute a shock." Gilbert had hit on the basic mechanism of the earthquake cycle, 30 years before it was laid out more fully by Harry Fielding Reid in the aftermath of the San Francisco earthquake (see Chapter 3).

Once Gilbert had conceived the basic idea of the earthquake cycle, he realized that earthquake prediction might be possible. He started with a "negative" prediction. When the residents of Lone Pine rebuilt their brick homes with lumber, the better to withstand a future earthquake, Gilbert declared that the extra expense was wasted, because the long-accumulating strain had already been released. "The spot which is the focus of an earthquake," he wrote, "is thereby exempted for a long time." We now know, of course, that this is a dangerous principle to live by, because a single segment of a fault can sometimes break twice within a few decades, or a neighboring fault may jump on the seismic bandwagon. But in this particular case, Gilbert was right: the Owens Valley has not experienced anything comparable to the 1872 earthquake in the 125 years since.

Gilbert also made at least one positive prediction, about a fault at the other, eastern extremity of the Basin and Range province. Gilbert found that a long system of normal faults runs north-south along the foot of the Wasatch Range, the imposing eastern backdrop to Salt Lake City and numerous smaller communities in northern and central Utah. It was apparent to Gilbert that the Wasatch Range was a typical basin range that was being created by repeated ruptures along the fault system. He realized, however, that the entire 200-mile extent of the Wasatch fault system did not break as a unit; rather, the system was divided into segments that broke more or less independently. He made this deduction from the apparent ages of the scarps left by the most recent ruptures: along some stretches the scarps seemed relatively youthful, whereas along other segments they were greatly eroded.

When Gilbert came to study the Wasatch Range in the neighborhood of Salt Lake City, he couldn't find any fault scarp at all. He spelled out the apparent significance of this observation in an article he wrote for the *Salt Lake Tribune* in 1883. "There is one place," he wrote, "where [the fault scarps] are conspicuous by their absence, and that place is close to this city. From Warm Springs to Emigration Canyon, fault scarps have not been found, and the rational explanation of their absence is that a very long time has elapsed since their last renewal. In this period the earth strain has been slowly increasing, and some day it will overcome the friction, lift the mountains a few feet, and reenact on a more fearful scale the catastrophy of Owens Valley."

Gilbert's prediction must have come as something of a surprise to the citizens of Salt Lake City. Still, they were not unduly perturbed, and in fact several generations have passed since 1883, without a rupture anywhere along the Wasatch fault system.

In recent years, a number of geologists, including Michael Machette and his colleagues at the USGS in Denver and Menlo Park, as well as scientists from the Utah Geological and Mineral Survey, have carried out much more detailed studies of the Wasatch fault zone. They confirm Gilbert's basic finding that the fault zone is divided into a number of segments—they count 10 in all—that tend to break independently. The boundaries between segments are marked, for the most part, by obvious geological landmarks that offer resistance to the propagation of a fault rupture. Between the Salt Lake City segment and the Provo segment to the south of it, for example, a large spur of bedrock, the Traverse Range Salient, juts several miles out from the front of the Wasatch range. The fault has to make a 5-mile dogleg to get around the obstacle.

Beginning in the late 1970s, Machette and his colleagues cut trenches across the scarps at dozens of locations along the fault zone. They estimated the size and dates of previous ruptures by methods similar to those used on the San Andreas fault, the difference being that the displacements are vertical rather than horizontal. Their results show that, on average, any given segment of the fault has ruptured about once every 2000 years. A rupture has occurred somewhere on the central part of the fault zone (the 200-mile-long stretch centered on Salt Lake City) about once every 400 years.

Although the Salt Lake City segment last broke in the more distant past than the segments immediately to its north and south, it is not the part of the fault that seems closest to its next rupture. That honor goes to the Brigham City segment, a 25-mile-long segment about 75 miles north of Salt Lake City. The Brigham City segment has not broken in the last 3500 years, but before that it broke three times over a period of about 1000 to 3000 years. Thus, this segment has been accumulating strain

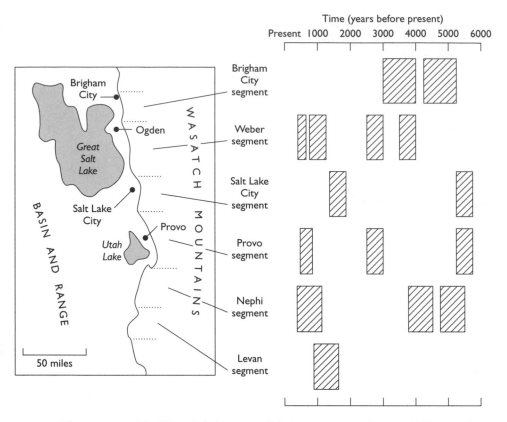

The segments of the Wasatch fault zone and their ruptures over the past 6000 years. In the diagram at right, the width of the blocks indicates the range of uncertainty for the date of each earthquake. (Adapted from illustrations by M. Machette and others, USGS)

energy for considerably longer than the average recurrence interval between ruptures. In that sense, another rupture on this segment could be described as overdue. Given the length of the segment and the amount by which it slipped on previous occasions, a rupture of the Brigham City segment would probably deliver an earthquake with a magnitude of around 7.1. Thus, it poses a serious hazard to Brigham City and the near-by communities. Unfortunately, the infrequency of rupture of the Wasatch fault makes mitigation efforts much more difficult to accomplish than they would be in such places as Los Angeles and San Francisco, where the Earth herself allies with the doomsayers to motivate action.

Just as with the San Andreas fault, however, the history of ruptures on the Wasatch fault zone is too irregular in space and time to make firm and reliable predictions about which segment is next in line for an earthquake. In particular, the ruptures of different segments have clustered together in time: five adjacent segments of the fault zone, for example, ruptured during an approximately 1000-year period that ended about 400 years ago. If such clustering is more than coincidental, it suggests that the ruptures of individual segments alters the probability of rupture on neighboring segments, just as has been proposed for the San Andreas and other faults in California. Thus, the most secure conclusion to be drawn from the research on the Wasatch fault zone is that the entire strip of land along the base of the mountains, where 80 percent of Utah's population lives, has a modest but significant likelihood of experiencing a major earthquake in the foreseeable future. To put a number on it, the official USGS assessment is that most of this region has a 10–20 percent probability of experiencing a ground acceleration of more than 0.2 g (about the level experienced by Los Angeles's San Fernando Valley during the 1994 Northridge earthquake) within the next 100 years (see Color Plate 18).

The Central Nevada Seismic Zone

There is another part of the Basin and Range that seems to be experiencing a series of clustered earthquakes right now. These earthquakes have been happening in west-central Nevada, in a little-visited region east of Reno and the Carson Sink. The series started at the beginning of the twentieth century or even earlier, but the events of greatest notoriety began in 1915. In that year, a rupture tore along 40 miles of the west side of the Tobin Range, south of Winnemucca and east of Reno. The rupture caused an earthquake of about magnitude 7.2 to 7.5 (the Pleasant Valley earthquake), and raised scarps up to 19 feet high. In 1932, a magnitude 6.7 earthquake struck Cedar Mountain, about 120 miles to the south of Pleasant Valley. In 1954, four earthquakes occurred within a 20-mile-wide area between the sites of the 1915 and 1932 earthquakes. The first two, in July and August, were in the Rainbow Mountains, east of Fallon; their magnitudes were 6.2 and 6.5. Then, on December 15, three faults ruptured more or less simultaneously in the vicinity of Fairview Peak, one range to the east of Rainbow Mountain. The earthquake produced by the collective rupture of these faults had a magnitude of 7.2. Just four minutes after the Fairview Peak earthquake, yet another earth-

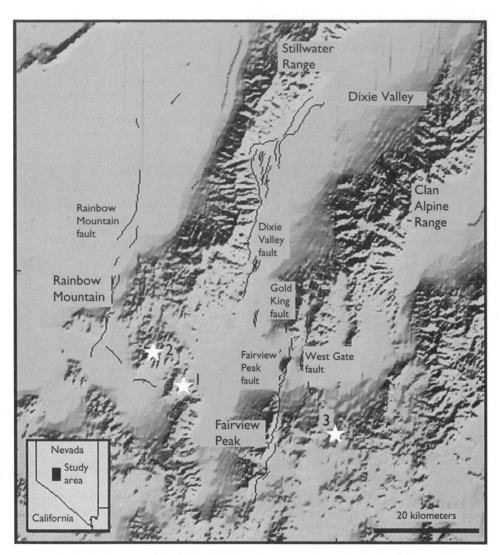

Shaded-relief map of the Dixie Valley area of western Nevada showing faults and the epicenters of the 1954 earthquake sequence (stars). (Courtesy of K. Hodgkinson and others)

quake (of magnitude 6.7) rocked the region: this one was caused by rupture of the Dixie Valley fault, which runs northward from the Fairview
Peak area toward (but not as far as) the fault that ruptured in 1915.

The close spacing of these earthquakes—especially those that happened in 1954—suggests that some kind of domino effect is at work. A
group from the USGS at Menlo Park and the Institut de Physique du

Globe in Paris (Kathleen Hodgkinson, Ross Stein, and Geoffrey King) examined the four 1954 earthquakes, using the technique of stress-transfer analysis that we described in Chapter 5: they calculated how each rupture changed the shear and clamping stresses on other faults nearby. The analysis was more complex than for the California faults, because most of the faults in western Nevada rupture in some combination of normal dip-slip and right-lateral strike-slip motion; therefore, stress changes had to be calculated in a three-dimensional volume, not just across a two-dimensional surface.

Hodgkinson and her colleagues found that stress changes induced by the first two earthquakes in the 1954 sequence (the two Rainbow Mountain earthquakes) brought the Fairview Peak faults and the Dixie Valley fault closer to failure. The effect was much stronger on the Dixie Valley fault than on the Fairview Peak faults, yet the Fairview Peak faults broke first. This may well have been because several millennia's worth of strain had built up on the Fairview fault, abetted by the 1932 Cedar Mountain rupture that took place just to the south of Fairview Mountain. Then the combined stress changes induced by the Rainbow Mountain and Fairview Mountain earthquakes were enough to bring the Dixie Valley fault to failure.

No large earthquake has occurred in this region since the 1954 cluster, but small earthquakes continue, generally in areas where stresses have been increased by the 1954 earthquakes. One particular area of interest is the 25-mile-long "seismic gap" that separates the northern end of the Dixie Valley rupture from the southern end of the fault segment that ruptured in the 1915 Pleasant Valley earthquake. According to Robert Wallace and Robert Whitney, there are no recent fault scarps in this gap, meaning that it last broke a long time (probably thousands of years) ago. They suggest that this region, known as the Stillwater gap, has been brought closer to failure by the 1915 and 1954 ruptures to the north and south and may be the next domino to fall. There is nothing, however, to indicate that an earthquake on the Stillwater gap is imminent.

Gilbert Revised

It is not uncommon in science that a theory starts off as a radical threat to established doctrine, is then gradually assimilated into the mainstream of accepted knowledge, and finally becomes a fossilized dogma itself, blocking further progress. Such was the fate of Gilbert's block-tilting theory for the creation of the Basin Ranges. Initially, Gilbert had to face conservative critics such as Josiah Spurr, who held to the notion that "monoclinal ridges are in most cases parts of anticlinal or synclinal folds;" in other words, that ramplike arrangements of strata are the remaining halves of arch- or trough-shaped folds that have somehow lost their other halves to erosion. By the time of his death in 1918, however, Gilbert was a world-renowned scientist, revered for his many brilliant contributions to geology. (For example, he was the first to realize that the moon's craters were the product of impacts, not volcanoes.) And block tilting had become the only legitimate way to think about the Basin Ranges; observations that didn't fit with that theory were either ignored or were shoehorned into conformity with it. Yet, as it turns out, block tilting is only a part, perhaps even a minor part, of the process by which the Basin and Range province has acquired its present structure.

One problematic finding in the 1970s was the observation of faults that couldn't easily be produced by block tilting. These were very gently sloping (in fact, almost horizontal) faults that underlay wide extents—hundreds of square miles—of the province. One such fault, for example, was found near Beatty, in the desert east of Death Valley. Faults of this kind are often called *detachments,* because they have detached an upper from a lower layer of the crust.

The interpretation initially advanced for these faults was that they were thrust faults: like the faults under Los Angeles, they had been wedging a lower block under an upper block, raising the surface and thickening the crust. Because the crust of the Basin and Range province is now extending, these thrust faults would have to be extinct remnants of a period, 20 million and more years ago, when contraction, not extension, was going on.

Long-extinct thrust faults do exist in the Basin and Range province, but the faults at Beatty and elsewhere are different. For one thing, where they come to the surface, they cut through the volcanic layers that were laid down at the beginning of the period of extension. So they must have been active after the period of compression was over.

Even more important, however, was the discovery, in the lower (footwall) face of these faults, of great masses of rock—sometimes entire mountain ranges—whose crystalline structure had been altered by exposure to high temperature and pressure. Such *metamorphic* rocks must have

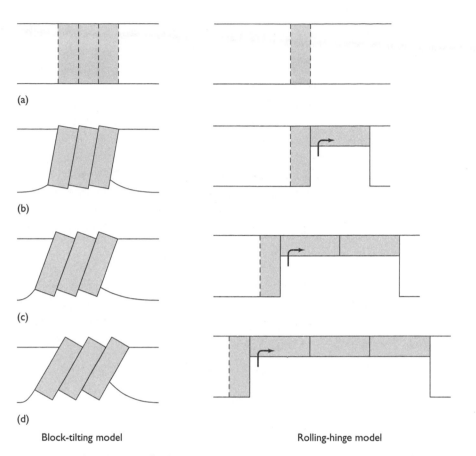

(a)

(b)

(c)

(d)

Block-tilting model   Rolling-hinge model

Two proposed mechanisms for the extension of the Basin and Range province. Both dia-grams are highly schematic; for example, the initial faults are probably oblique rather than vertical in both cases.

spent some time deep in the crust—a good 8 or 10 miles below the sur-face, at least. Yet classical block tilting can't expose such deep rocks, any more than tilting a row of books can bring the lower parts of the books to the top surface. Furthermore, it proved possible, by measuring the ratios of various radioactive isotopes in the metamorphic rocks, to estab-lish when they were brought up out of that pressure-cooker environ-ment. It turned out that they were brought up less than 15 million years ago—at or after the beginning of the period of extension. So they weren't a mere remnant of some process that had churned up the crust *before* extension began; rather, their rise to the surface seems to have been an integral part of the process of extension itself.

One of the scientists who has made a particular study of the detach-ment faults is Brian Wernicke of Caltech. Wernicke believes that the Basin and Range province can be divided up into alternating bands, each band several ranges wide. In one kind of band, extension has fol-lowed the classic block–tilting scheme. These regions have extended fairly modestly—a total of 10 to 15 percent lengthening over the past 15 million years—and their surface is more or less the same surface that existed prior to extension. The other set of bands, however, has expand-ed much more, and by a radically different process that Wernicke terms the "rolling-hinge mechanism." In these regions, the normal faults not only have cut the crust into oblique slices, but the slices have then been dragged in sequence across the surface, like a series of slices of freshly cut ham being moved along a conveyer belt. The surface of these regions is therefore mostly "new land" formed from the fault surfaces that once were situated in the deep interior of the crust. When Wernicke and his colleagues estimated how much extension had occurred in these regions, they concluded that the entire Basin and Range province had at least doubled in width since the process of extension began.

Wernicke believes that most of the latter, large-scale extension hap-pened relatively early in the province's evolution. The process of exten-sion, he believes, has slowed down from an original rate of about 20 millimeters (0.8 inch) per year to the current rate of about 10 millime-ters (0.4 inch) per year, and most extension at present is happening by the classic block–tilting process.

But the rapid, rolling-hinge process may not be completely extinct. Wernicke is particularly interested in one detachment that has been detected by seismic soundings under Utah's Sevier Desert, about 80 miles south of the Great Salt Lake. He believes that this detachment may well be a still-active low-angle normal fault, and he is currently setting up an experiment to measure strain accumulation over the area, using the GPS technique. Of course, measuring these tiny deformations will take several years of patient observation ("That's why we need tenure," Wernicke joked recently.). If he should find the deformation pattern that he is looking for, it would be very significant for the residents of south-western Utah. Such a discovery would show that they are exposed to the danger of an earthquake that, on account of the great area of the fault plane that could rupture at one time, might be greater than anything that has struck the Basin and Range province in human history.

The Basin and Range province is not the only part of the Earth's surface that is stretching. A very comparable region is the Aegean Sea. The Greek islands are moving slowly to the southwest, away from Asia Minor. As a result, the floor of the Aegean has become a series of tilted blocks whose structure is very similar to that of Nevada and Utah. The big difference, of course, is that the Aegean is largely under water. Those picturesque islands, with their whitewashed villages and cloth-sailed windmills, are the scattered peaks of basin ranges produced by normal faulting.

So where does our journey take us next? Not so far, really. For just as one active volcanic area, the Long Valley caldera, marks the western margin of the Basin and Range province, so another, the Yellowstone caldera, lies on its eastern edge. One of the world's most productive volcanoes, Yellowstone is driven by a geological process wholly different from those that produce the volcanoes of Cascadia or Long Valley—a process that has its roots much deeper in the mantle, more than 1000 miles beneath the Earth's surface.

9

ON THE HOT SPOT: THE
VOLCANOES AT
YELLOWSTONE

Old Faithful *used* to be faithful. Once an hour, more or less, the most celebrated geyser in North America flung its jet of scalding water skyward, to a chorus of "oohs" and "aahs" from the assembled onlookers. The geyser's predictability was an icon at Yellowstone National Park. But the Hebgen Lake earthquake changed that, by throwing some cogwheel in the geyser's inner clockwork off its bearings. And ever since August 17, 1959, it has erupted less often and less regularly.

The changes in Old Faithful—and at hundreds of other geysers, hot springs, mudpots, and fumaroles—caused by the Hebgen Lake earth-

An eruption of Old Faithful geyser, Yellowstone National Park. Note the pulses within the jet. (Photo by S. W. Kieffer)

quake were a gentle, even trivial reminder of how impermanent the Yellowstone landscape really is. Over the centuries, the geysers come and go. Over the thousands of centuries, the volcanoes themselves, the powerhouses that fuel the park's whole watery spectacle, also come and go. At times, Yellowstone has lain buried under thousands of feet of ice. At other times, its fiery eruptions have extinguished all life for hundreds of miles around.

Ten million years ago, Yellowstone was not volcanic at all. Now, although it may not seem that way, the park is in the throes of intense volcanism. And 10 million years in the future, the whole region may well be a quiet and level plain, with little to suggest—except to a geologist—that volcanoes ever disturbed its tranquillity.

Hot Spots

At about the same time that G. K. Gilbert was exploring the Basin and Range province, another geologist, Ferdinand Hayden, was making the first scientific observations at Yellowstone. Hayden realized that the entire Yellowstone basin is a volcanic caldera, the remains of an immense eruption that took place at some time in the distant past. Decades later, geologists recognized that there are three distinct but overlapping calderas in the Yellowstone region, and they mapped the ashfall deposits left by their three climactic eruptions—deposits that reach from the Pacific Ocean to the Gulf of Mexico. In the 1960s, with the advent of the potassium-argon dating method, the eruptions could be given dates—2.0 million, 1.3 million, and 0.62 million years ago. Yet, for all these scientific advances, there was no understanding of why Yellowstone has been so singularly plagued with volcanoes. This understanding awaited the late 1960s and the theory of plate tectonics.

In Chapter 1, we described some of the evidence that led to the theory: the apparent fit between the margins of some neighboring continents; the discovery of mid-ocean ridges, subduction zones, and transform faults; and the detection of the magnetic stripes that record the history of the widening ocean floors. According to the theory, which more recently has been confirmed by GPS and other space-borne surveying methods, large sectors of the lithosphere are in slow but constant motion, each along its own path and at its own speed. The motions are relative: you can say that London (on the Eurasian Plate) is stationary and New York (on the North American Plate) is moving west, or you can say that New York is stationary and London is moving east. Even the mid-ocean ridges, which at first one might think of as stationary, are in fact mobile. This becomes obvious if, for example, one considers the fact that there is no subduction zone between the ridges in the Atlantic and Indian oceans; therefore, the African Plate must be getting larger and the two ridges must be moving apart. Which ridge is actually moving, however, the theory of plate tectonics does not presume to say.

Yet curiously, just as our notion of a fixed Earth was dissolving into a tectonic free-for-all, a new and completely unexpected frame of reference was discovered: a set of fixed landmarks deep in the Earth's mantle, against which the motions of the overlying lithosphere could be measured. The discovery of these markers—the so-called *hot spots*—played a crucial role in the early work on plate motion.

The concept of hot spots was first put forward by the Canadian geophysicist J. Tuzo Wilson in 1963 to explain the origin of linear chains of oceanic volcanoes, such as those of the Hawaiian Islands. Wilson suggested that the chains were formed as the drifting lithosphere passed over a fixed source of heat in the underlying mantle; the volcanoes were like the scorch marks left on a sheet of paper that is moved across a candle flame. (We shall have much more to say about the specific origin of the Hawaiian volcanoes in Chapter 12.)

W. Jason Morgan, whose contribution to plate tectonics was mentioned in Chapter 1, strengthened Wilson's hypothesis in two ways. First, he and his colleagues figured out a set of motions for the plates that was internally consistent—that is, the relative motion between all neighboring plates corresponded to the magnetic and other data—and that gave a motion for each plate that was equal and opposite to the apparent motion of the hot spots on that plate. In other words, if one allowed for the motion of the plates, then the hot spots were stationary, or nearly so. It was truly the paper being moved across the candle, and not the candle beneath the paper.

Morgan's second contribution was to suggest an actual physical explanation for hot spots. He put forward the idea that hot spots are the surface expression of *mantle plumes*—masses of rock that ascend from the deeper parts of the mantle under the influence of convection. Having reached the underside of the lithosphere, they heat it sufficiently to induce volcanism.

Mantle plumes are difficult to observe directly, and much of what we "know" about them is simply the most plausible of various scenarios, given our general knowledge of conditions in Earth's interior. But a consensus account of them runs something like this: Heat continuously radiating out of Earth's core gradually raises the temperature of the deepest layer of the mantle. This layer, like nearly all the mantle rock, is solid—kept that way, despite its high temperature, by the pressure of the overlying mantle and crust. But it is sufficiently ductile that, given enough time, it can flow like a liquid. As its temperature rises, it expands, becoming less dense than the overlying, cooler rock and therefore ready to rise through it. Sooner or later, therefore, but with almost unimaginable slowness, a great blob of hot rock heaves itself away from the core-mantle boundary and begins its ascent toward the surface. The blob of rock may have a volume of about a million cubic miles (equivalent to a sphere 120 miles in diameter), and one such blob may leave the core-mantle interface every few million years.

As the plume head ascends through the mantle, a much narrower conduit—only 10 to 20 miles or so across—trails behind it, so that the whole plume resembles a balloon trailing a string. Hot rock ascends both in the head and in the trailing conduit. This kind of behavior can be replicated in the laboratory: if hot, colored glycerin is injected continuously into the bottom of a tank full of cooler, clear glycerin, the hot glycerin rises in the form of a balloon followed by a string.

When the head of the plume reaches the undersurface of the lithosphere, the excitement begins. Rock in the plume head, as well as some of the overlying rock that has been heated by the plume, partially melts and forces its way up through the crust, leading to volcanic eruptions that cover wide areas of Earth's surface with basalt. On the continents, the end products of such volcanism are *flood-basalt provinces,* which may extend for many hundred of miles. About 20 such provinces, the residues of plume heads that reached Earth's surface in the ancient past, have been mapped. Plume heads produce similar basaltic outflows when they rise under oceanic crust. The resulting provinces, however, are less obvious, not only because they are submerged but also because the ocean floor consists of basalt anyway.

The scale of the volcanism associated with some of the flood-basalt provinces is truly impressive. One such province, the so-called Deccan Traps, covers most of northwestern India. And we are talking here, not of ash carried by the wind, but of lava flows thousands of feet thick. The

Deccan Traps erupted about 65 million years ago, at just about the time the dinosaurs (as well as the majority of all the species living at the time) became extinct. Although most geologists now believe that a meteorite impact was the prime cause of that extinction event, the Deccan Traps volcanism may have played a contributory role, for eruptions on that scale would very likely have had major effects on the world's climate. Other flood-basalt provinces are found in Siberia, Brazil, and southern Africa.

Flood-basalt volcanism, though intense, is mercifully brief: within a mere million years of the plume head's reaching the lithosphere, its heat has dissipated. Even a fast-moving plate will travel no more than 100 miles in this time, so the "scorch-mark" created by the broad plume head is not noticeably elongated in the direction of the plate's motion.

The story of the narrow conduit that follows the plume head to the surface is quite different. The geological evidence is that the conduit may remain in existence for 100 million, sometimes even 200 million, years after the plume head has dissipated, and all that time it is conveying hot rock from the deep mantle to the lithosphere. Furthermore, the conduit is tiny—just a few miles across. So its effect is not to create a broad volcanic province but to inscribe a narrow ribbon of volcanoes on the plate that is tracking across it. This narrow ribbon is particularly obvious if the overlying crust is oceanic; that is, thin and uniform. That is why the Hawaiian Islands and other oceanic volcanic chains are distinct and linear. When the conduit is under the thicker and more variegated crust of a continent, the hot-spot track may be less obvious, but it is there nonetheless.

Eventually, without any special closing fireworks, the conduit also dissipates. Thus the hot-spot tracks, which start at flood-basalt provinces, end either at a currently volcanic region (if the conduit is still in existence) or at a region whose volcanism has ceased (if the conduit has dissipated).

Robert Duncan of Oregon State University at Corvallis, along with a number of other scientists, took on the challenge of mapping the world's hot-spot tracks. Imagine their efforts as a board game: select a flood-basalt province, select a hot spot, and see if you can join them with a chain of extinct volcanoes whose ages decrease systematically from the flood-basalt province to the hot spot. If you succeed, select another flood-basalt province; if not, it's someone else's turn. Connecting the dots would be a trivial task if the tracks were always

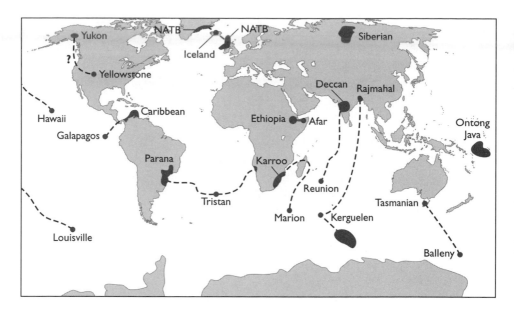

Map of the Earth's flood-basalt provinces, hot-spot tracks, and current hot spots. NATB = North Atlantic Tertiary basalts. No track has yet been associated with the ancient Siberian flood-basalt province. (Adapted from an illustration by R. Duncan and M. Richards)

straight and uninterrupted. But they are straight only if the plate over a given mantle plume remained in constant motion in the same direction over the entire life of the hot spot. That is unusual; more often, the plate changes direction at some point, so that the hot-spot track is left with at least one bend. As we'll see in Chapter 12, the Hawaiian hot-spot track has two such bends.

Even more confusingly, hot-spot tracks may be interrupted by big gaps. This happens when a plume track has been overridden at some time in the past by a mid-ocean ridge. An example of such a track is the one that starts at the Deccan Traps. It runs south for a couple of thousand miles into the Indian Ocean and then stops dead. The remainder of the track picks up about 600 miles westward, to the east of Madagascar, and then runs southwest to its termination in a currently active hot spot near the island of Reunion. (The Piton de la Fournaise volcano on Reunion erupted as recently as March 1998.)

This pattern is explained as follows. When the plume head first reached the lithosphere, 65 million years ago, northwest India was where Reunion is now. Since then, India has drifted north toward its fateful collision with Eurasia. The hot-spot track would therefore run

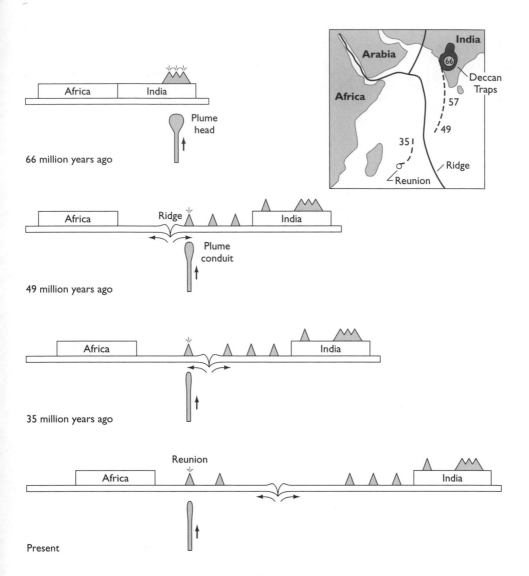

The history of the Reunion hot spot. The inset shows the present track, running from the Deccan Traps flood-basalt province in northwest India to the island of Reunion, east of Madagascar. The numbers give the ages of volcanic rock from various sites along the track, in millions of years, as determined by potassium-argon dating. The four diagrams show four stages in the history of the track, from its birth 66 million years ago (m.y.a.) to the present. Note that the mid-ocean ridge drifted eastward across the hot spot about 37 million years ago. Since that time, the continuous creation of new ocean floor at the ridge has been opening up a gap in the track.

directly from India to Reunion, except for the fact that, about 37 million years ago, the ridge that runs down the middle of the Indian Ocean drifted eastward across the track. (Remember, we mentioned that the locations of the ridges are not fixed with respect to the Earth's interior.) Thus, for the past 37 million years, new oceanic crust, unscarred by any hot spot, has been opening up a gap in the track.

Just to complete the rules of the game, we should mention what happens when a hot spot is located at a mid-ocean ridge that is stationary, or nearly so. This is the unfortunate situation of Iceland, the large volcanic island that sits directly astride the Mid-Atlantic Ridge, between Greenland and northern Europe. The flood-basalt province that began the Icelandic hot-spot saga, 60 million years ago (labeled "NATB" in the illustration on page 187), was torn in two by the opening of the North Atlantic: half of it now lies in eastern Greenland, and half lies in Scotland and nearby. Thus, two hot-spot tracks lead from these two provinces to a single hot spot at Iceland. For the same reason, double tracks lead to the Tristan da Cunha hot spot in the South Atlantic and to the Kerguelen hot spot in the Indian Ocean.

By skillfully playing this game over a number of years, Duncan and other scientists not only have constructed a map of the most of the world's hot-spot tracks, but also have been able to measure how stable the array of hot spots really is. It turns out that the array of hot spots is stable to within the limits of the measurement technique, which is about 3 millimeters (0.1 inch) per year. If individual hot spots drift sideways at all, they do so no faster than this. In a sense, then, the array of hot spots in the mantle beneath our feet is more fixed than the array of stars in the sky over our heads, for many stars are moving detectably with respect to other stars.

Yet Duncan, along with R. B. Hargraves, capped this beautiful piece of analysis with another, unexpected finding. The entire array of hot spots—and thus the entire mantle—seems to be moving with respect to yet another reference frame: the frame established by Earth's axis of rotation. The way they figured this out was by measuring the magnetism of volcanic rocks at various sites along hot-spot tracks. Because Earth's magnetic lines of force emerge from the planet's surface at different angles ("inclinations") at different latitudes, the orientation of the magnetism locked into the volcanic rock at its time of cooling tells us the latitude of each volcano at the time when it was erupting. Thus, if hot spots are stationary with respect to the north and south magnetic poles, the recorded latitudes should be identical for all sites along a given track. But they are not. Instead, they change systematically along the track. Analysis of these changes along many hot-spot tracks shows rather clearly that the mantle is slowly rolling with respect to the magnetic poles. Because the magnetic poles always stay near the poles established by Earth's axis of rotation (the "geographic poles"), it can be concluded that the mantle is rolling with respect to the rotation axis also. Specifically, if a rather long toothpick were driven through the Earth, starting on the equator near the Seychelles in the Indian Ocean and continuing through to the equator on the other side of the globe (in the Pacific Ocean), then the mantle is presently rotating clockwise (viewed from the Indian Ocean side) around this toothpick at a rate of about 3 degrees every 10 million years. Over the course of geological time, however, the direction and speed of rotation has changed erratically. It is believed that these rolling motions of the mantle reflect the redistribution of mass within the Earth that is caused, for example, by continental drift and by convection within the mantle.

The Yellowstone Track

Yellowstone is a spot that is hot—that much would be obvious to anyone who dipped a toe in Mammoth Hot Springs—but is it a hot spot? What is the evidence, in other words, that volcanism at Yellowstone results from the interaction between a mantle plume and the overlying lithosphere, rather than from some other process? The evidence is of two general kinds: the discovery of what looks like a hot-spot track leading to Yellowstone, and the finding of conditions at Yellowstone that are consistent with the presence of a mantle plume there now.

The evidence for the hot-spot track is very persuasive. We know that the North American Plate is currently drifting southwest (with respect to the mantle) at a rate of about 30 millimeters (1.2 inches) per year. Therefore, if there is a hot-spot track leading to Yellowstone, it should come from that direction, and any volcanoes that form part of the track should have been active at progressively earlier times the farther to the southwest they are located.

This is exactly what has been found. To the southwest of Yellowstone lies the broad, flat Snake River Plain (see Color Plate 19). About 60 miles wide and extending across nearly the entirety of southern Idaho, the plain sweeps through the northern part of the Basin and Range province, its flatness offering a marked contrast to the rugged terrain to the north and south. The flatness of the Snake River Plain is a result of outpourings of thin, basaltic magma from the many volcanic vents in the region. These outpourings obliterated most of the traces of earlier, siliceous volcanism, but some of these earlier volcanoes, such as Big Southern Butte at the northern edge of the plain, still raise their heads above the stark black sea of basalt.

In the 1970s and 1980s, geologists mapped and dated the remnants of a string of volcanic edifices and calderas, some as big as those at Yellowstone, that lie partially exposed at various sites along the plain. Here are some of the results. At the northeast tip of the plain, about 100 miles southwest of Yellowstone, the siliceous lava flows are about 6 million years old. Near Twin Falls, Idaho (about 250 miles southwest of Yellowstone), they are about 10 million years old. By the time one gets to the Idaho-Oregon border, 400 miles southwest of Yellowstone, the lava flows are 13 million years old. The sequence of ages is in the expected direction, and after one has corrected for the stretching of the Basin and Range province, the dates correspond roughly with what would be expected, assuming a plausible rate of movement of the North American Plate during this time. Thus, the plate very likely is drifting across a stationary source of heat.

There is much more evidence for a hot-spot track, however, than just the volcanoes and their ages. For one thing, seismic reflection studies suggest that the 5- to 10-mile-thick layer of granite that forms the "basement" for much of the North American continent is absent or unusually thin under the Snake River Plain. It has been replaced by rock that conducts seismic waves faster than granite and is therefore denser. Also, the force of gravity is slightly stronger in the Snake River Plain than in the surrounding region, again suggesting that there is unusually dense rock in the upper crust. Most probably, this is rock from the upper mantle that melted under the influence of the passing plume, rose into the crust, and hardened there.

Another piece of evidence comes from the pattern of earthquakes in the region. Yellowstone itself has many earthquakes, of course. But looking at the wider region, one can see that most historic earthquakes have occurred in two "wings" that converge on Yellowstone from the west and the south-southwest. They have taken place, in fact, among high ranges to the north and south of the Snake River Plain. The earthquakes have been caused by typical basin and range–style ruptures, for the most part. The Borah Peak and Hebgen Lake earthquakes, for example, both took place within the northern wing (see the Introduction and Chapter 8). What is unusual is not the style of the earthquakes, but their number.

The "wings" demarcated by the earthquakes look a bit like the pattern of spreading waves in a stream as it flows past a rock. The "rock," of course, is Yellowstone itself, which currently lies right over the plume conduit. After the hot material in the conduit reaches the lithosphere, it is dragged southwestward by the motion of the overlying plate, rather as smoke ascending in a factory chimney is swept downwind once it exits the top of the stack. But as the plume material moves southwestward, it also spreads out sideways in the pattern of the expanding wings. The heat causes the crust in the wing regions to expand and rise, which results in earthquakes, mostly along normal faults. The centerline of the track—the Snake River Plain—is sinking because it is cooling (as well as because of the increased density of the crustal rocks there), and it is seismically quiet.

The Snake River Plain, then, looks very much like a hot-spot track. But if one follows it westward, looking for its origin, things become fuzzy and contentious. In southern Oregon and northern Nevada there are the remains of a large number of calderas dating from 14 to 17 million years ago. Very likely, these calderas owe their existence to the Yellowstone hot spot. According to some geologists, the broad extent of volcanism in this area means that this region was over the head of the mantle plume when it first reached the lithosphere. Most volcanic eruptions in this area, however, produced high-silica rhyolite, not the flood basalts that are supposedly characteristic of the beginning of a hot-spot track.

To the north of this volcanic province, on the other hand, there *is* a flood-basalt province. Most of the high country of eastern Oregon, as well as the Columbia River Basin in northern Oregon and southeast Washington, was once inundated in basalt: fifty thousand cubic miles of it came pouring out of dikes, obliterating the preexisting topography. The eruptions happened between 17 and 13 million years ago, and most of the basalt was erupted within a much briefer time span—less than 2 million years.

Up until about 1980, the Columbia River Basin–Oregon Plateau flood-basalt province was considered a likely candidate for the origin of the Yellowstone hot-spot track. But there were problems. First, big though it is, the province is small in comparison with most flood-basalt provinces that have been identified as the origin of hot-spot tracks. The Deccan Traps province, for example, is an order of magnitude larger. Second, it's not quite in the right place: it's centered 250 miles north of the region that's thought to have been over the mantle plume 15 million years ago. And third, there are indications of even older traces of the hot-spot track farther to the west; if those are indeed part of the same track, then the Columbia River Basin province is not only too far north but also too far east, and too young, to have been produced by the ascent of the plume head.

Robert Duncan now believes that the Columbia River Basin province was some kind of sideshow laid on by the plume conduit as a huge obstacle drifted over it: the subducting slab of the Farallon Plate (the eastern half of the Pacific seafloor, now reduced to a few remnants, such as the Juan de Fuca Plate). We last met this plate under the Cascadia range, where it is currently provoking volcanism (see Chapter 2). But the sinking plate extends even farther to the east under Washington and Oregon. As this part of the plate passed over the plume conduit, Duncan suggests, it may have temporarily blocked the plume's ascent and then deflected the accumulated mass of plume material—a kind of "mini" plume head—northward to the region of the Columbia River Basin, where it gave rise to a "mini" flood-basalt province.

What seem to be older traces of the hot-spot track have been found in the coastal ranges of Oregon, 650 miles west of Yellowstone. Volcanic rocks there date to about 50–60 million years ago, and certain details of the rock chemistry are believed to be characteristic of plume-derived magmas.

Where, then, is the starting point of the Yellowstone hot-spot track? That question remains an enigma. A major problem with answering it is the fact that the history of the western edge of North America over the last 100 million years is extraordinarily complex. Fragments of continents have drifted in on the Pacific Plate, fusing themselves to the mainland. Other pieces that originally lay in the neighborhood of California and Oregon have migrated north, ending up in Canada or Alaska. Figuring out what piece of contemporary North America, if any, lay over the Yellowstone mantle plume more than 60 million years ago is no easy task. But according to one recent study by Stephen Johnston of the Yukon Geoscience Office, and several colleagues, a group of basaltic deposits in southwestern Yukon are the eroded remnants of a flood-basalt province that once covered much of the territory. These rocks—the Carmacks Group—were laid down 70 million years ago, and, according to the magnetic inclinations locked in the rock, they have migrated about 1200 miles northward since they were formed. Thus, they are a candidate to be the remnants of the flood-basalt province that marked the initiation of the Yellowstone hot-spot track.

Yellowstone's Volcanic History

Since the region of the crust that is now Yellowstone arrived over the mantle plume, 2 to 3 million years ago, the region has been one of the world's most active volcanic centers. The sequence of eruptions has been the particular study of Robert Christiansen of the USGS at Menlo Park. The first of the three giant eruptions, which took place 2 million years ago, left a caldera measuring about 50 by 30 miles; its exact size is hard to determine because it has been so battered by later eruptions, but it occupied the southwest quadrant of the present park and extended well into eastern Idaho. The volume of material ejected was about 600 cubic miles (measured as magma before it expanded during the eruption)—four times more than the Bishop Tuff eruption at Mammoth and over a *thousand* times more than the volume ejected by Mount St. Helens in 1980. The resulting "Huckleberry Ridge ash" can be found not just in the neighborhood of Yellowstone, but up and down the west coast of the United States, in northern Mexico, in Texas, and almost to the Mississippi (see the illustration on page 144). To spread so far, the Plinian column must have risen to an extraordinary height in the atmosphere and must have been sustained long enough for the wind to have changed direction several times.

The second caldera-forming eruption took place about 1.3 million years ago at the western end of the earlier caldera, at Island Park, Idaho. Although this was the smallest of the three caldera-forming eruptions, its output—about 70 cubic miles of magma—was still larger than the combined output of all the world's volcanoes over the past 200 years. The third eruption, which took place 620,000 years ago, ejected about 250 cubic miles of material (the "Lava Creek ash"), which spread from California to the coast of Texas. The resulting caldera (the Yellowstone caldera as we know it today) measures about 30 by 45 miles and over-

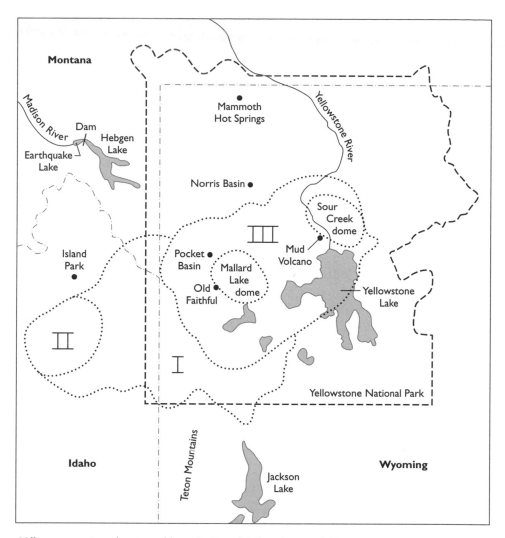

Yellowstone region, showing calderas (I, II, and III) and geyser fields.

laps the eastern half of the first caldera; its floor forms Yellowstone's Central Plateau and underlies most of Yellowstone Lake.

Adding in the smaller eruptions that took place between the major caldera-forming eruptions, the material erupted since Yellowstone came to lie over the hot spot has a total estimated volume of 2000 cubic miles. The sheer volume of volcanic output at Yellowstone supports the idea that a unique and long-lasting source of heat, such as a mantle plume, is driving the whole performance.

The Lava Creek eruption was not the end of Yellowstone's volcanic history. Since then, two resurgent domes have risen in Yellowstone caldera: first the Mallard Lake dome, in the western half of the caldera, and then the Sour Creek dome, to the east. Christiansen and his colleagues have found evidence for at least 40 distinct lesser silicic eruptions since the Sour Creek dome began forming, 150,000 years ago, as well as a number of basaltic eruptions near the edge of the caldera. The most recent eruption took place 70,000 years ago.

Ice

Since the eruption that formed Yellowstone caldera 600,000 years ago, at least three ice ages have left their mark on the region. At least three times, ice caps covered the highlands, feeding glaciers that bulldozed the rugged volcanic terrain into the more gently rolling topography that we see today. Yet the volcano fought back: between the second and third glaciation, for example, large eruptions of rhyolite in the western part of the caldera blocked the path that the earlier glaciers had taken. Thus, during the third and most recent glaciation, only 12,000 to 20,000 years ago, the ice was diverted into a huge stream that ran northward into Montana, along what is now the valley of the Yellowstone River. This glacier, together with the ice fields that fed it, covered 1300 square miles and averaged more than 2000 feet in thickness. At some spots, such as the Mud Volcano area to the west of the Sour Creek dome, the ice was more than 3000 feet thick.

The great weight of the ice influenced the behavior of the underlying volcanic system. Much of the volcano's heat is brought to the surface by the geysers and other hydrothermal systems in the form of hot water. When, as today, there is no overlying ice, the water that emerges at the surface can be no hotter that the boiling point of water at the altitude of Yellowstone; that is, about 92°C. An overlying ice cap, however, acts like the lid of a pressure cooker, allowing the water to reach the rock surface at much higher temperatures. Thus, the paradoxical effect of the ice caps was to cause the subsurface rocks to heat up. When the glaciers retreated, it was as if the lid of the pressure cooker had been removed. Sometimes, the pressure dropped suddenly; this happened when lakes of hot water, trapped under the melting ice, burst abruptly out of their confines. Then the superheated water in the underlying

rock would flash to steam, blasting out a large crater. The sunken mead-
ow known as Pocket Basin, near the Firehole River in the Lower
Geyser basin, is an example of such a hydrothermal explosion crater.

Yellowstone Today

One piece of evidence that Yellowstone is still active is the high rate at
which heat is coming out of the ground. Averaged over the whole
caldera, the heat flow measures about 2 watts per square meter, which
is about 30 times the rate of heat flow in other parts of the North
American continent and nearly 3 times the rate of heat flow from the
Long Valley caldera. The heat flow from the entire Yellowstone caldera
is about 5 gigawatts—enough, if it could be converted into an equivalent amount of electricity, to supply the power needs of a city of 5 million people.

The heat is being brought up by two mechanisms—first, by the
ascent of hot water in the hydrothermal systems; second, by direct conduction through the subsurface rocks. The waterborne heat is distributed irregularly, of course; the geysers and hot springs are grouped in
certain basins, mostly near the resurgent domes or near faults that lead
away from them. The flow of conductive heat, on the other hand, is
rather uniform across the caldera, but it drops off very sharply at the
caldera's boundary. Robert Smith and his colleagues at the University
of Utah lowered temperature probes into the bed of Yellowstone Lake,
which straddles the edge of the caldera. In the parts of the lakebed that
are within the caldera, about eight times more heat was flowing out of
the underlying rocks than was flowing out of the parts that lay outside.
Such a sharp heat-flow boundary implies that the heat source is shallow—not more than a few miles below the surface.

Even more direct evidence that there is something going on under
the Yellowstone caldera comes from measurements of ground deformation. In the late 1970s, it was found that the floor of the entire caldera
was rising. Benchmarks first surveyed in the 1920s had risen by as much
as 30 inches. By the mid-1980s, a further uplift of 10 inches had been
detected. Then, in 1985, the process abruptly changed direction: by
1990, the floor of the caldera had sunk by about 5 inches. Just as at
Mammoth, the movements of the caldera floor have been accompanied
by swarms of small earthquakes. The earthquakes have mostly been very

shallow—less than 3 miles below the surface—suggesting that beyond that depth, the rock is too hot to sustain brittle fracture.

The exact cause of the elevation changes, and their significance in terms of the potential for future eruptions, is unclear. During the period of uplift before 1985, something—probably magma, but possibly hot brine—was migrating into the upper crust. Then, during the subsequent period of subsidence, the material that had intruded either spread sideways or lost volume, perhaps by the release of entrapped gases.

There is plenty of evidence that, since the last ice age, the Yellowstone caldera has undergone numerous cycles of uplift and subsidence. Ancient shorelines, both above and below the present level, show that Yellowstone Lake periodically extended several miles northward down what is now the Yellowstone River. The advances were caused by uplift of the caldera floor north of the lake, which periodically dammed the lake's outflow. Ken Pierce, of the USGS at Golden, Colorado, has analyzed the cycles of uplift and subsidence by carbon dating plant material that he found in the abandoned shoreline terraces. According to Pierce, the current cycle is quite modest in size and duration, compared to some episodes during the past few thousand years. Since those periods of uplift did not culminate in eruptions, it's reasonable to believe that the current period will also end peacefully.

How Geysers Work

Of all the signs of volcanism at Yellowstone, the geysers are most famous. Generations of visitors have been curious about their inner workings. Why do they jet briefly skyward rather than flowing calmly and continuously as most of the park's hot springs do? Why is there some kind of regularity to their eruptions? The basic answers to these question were provided by Don White, who studied the geyser at Steamboat Springs, near Reno, Nevada, in the 1960s. Later, he and others extended their observations to the geysers at Yellowstone.

The water that supplies Old Faithful and the other geysers and springs comes originally from the snow that falls on the mountains surrounding the Yellowstone caldera. The meltwater percolates into the rock under the caldera, where it is heated by the underlying magma to a temperature of about 400°C. The water does not contact the magma directly; rather, it is thought that the magma is capped by rock that is permeated with a very concentrated salt solution. Convection within this hot brine brings the magmatic heat up to the level of the groundwater, a few hundred feet below the surface. The heated water then rises through the volcanic rock, dissolving some of the rock's minerals, such as silica, along the way.

As the water rises and cools, the silica tends to come out of solution again, and it is deposited on the walls of the channels through which the water is rising. Over time, the buildup of silica creates partial blockages at various levels in the system, and these blockages act as throttles, causing the water below them to become more highly pressurized than it would be if it could flow freely to the surface. From time to time, earthquakes open new passages to the surface. The Hegben Lake earthquake, for example, may have diverted some of the water that previously emerged through the Old Faithful geyser into a different conduit. Something similar probably happened after the 1983 Borah Peak earthquake, which was also followed by a reduction in the size and frequency of the geyser's eruptions.

Eventually the water, now at a temperature of about 215°C, enters a reservoir at the bottom of the geyser's conduit. In the case of Old Faithful, this reservoir is about 70 feet below the surface. Susan Kieffer, while a USGS geologist at Flagstaff, explored the conduit with a remote-control video camera: rather than being a uniform-diameter pipe, the conduit has several cavities or enlargements along its course.

Right after an eruption, the geyser's conduit is empty, and the reservoir at the bottom is partially empty. Hot water from below steadily refills the reservoir over a few tens of minutes, and once the reservoir is full, the hot water begins to rise up the conduit. Initially, the rising water is cooled by contact with the relatively cold walls of the conduit, and it does not boil. But convection replaces the cooled water with more hot water from the depths, and heat is also carried upward by bubbles of steam that collapse in the cooler upper reaches of the column. Eventually, every point in the upper portion of the column of water is at the boiling point for its depth and pressure; in other words, the temperature is 93°C (the boiling point at the elevation of Yellowstone) at the top of the column and progressively hotter toward the reservoir.

Steam now begins to rise, not as bubbles but as large packets the width of the conduit that lift the upper part of the column some distance before collapsing or slipping past the overlying water. Driven by these packets of steam, the top of the column begins an irregular surging, and eventually the surges spill out of the top of the conduit in a sequence of low spurts that Kieffer calls preplay. (Geyser talk is rife with sexual allusions, of course; Kieffer's terminology seems like a riff on Masters and Johnson.)

For a few minutes, the preplay surges actually hold an eruption at bay, because they release some of the heat from the system. But eventually, the slugs of ejected water become so large that, as they leave the geyser's orifice, the pressure on the underlying water drops significantly. Therefore, some of the water flashes to steam, ejecting a much larger packet. This begins the initiation phase, a sequence of about five spurts that grow to the geyser's full height, which is about 130 feet.

A chain reaction of decompression now rushes down the conduit and into the reservoir, causing water to flash to steam at ever greater depths. This is the steady flow stage, about 30 seconds long, during which the spout of ejected water stays near its maximum height of about 130 feet. When carefully observed, however, the "steady" flow is not steady at all, but consists of individual pulses, each lasting about half a second. In a still photograph of the spouting geyser, the fragmentation of the jet into about 8 to 10 pulses can be readily discerned. The reason for the pulses is not completely understood, but it suggests that the underground conduit acts as a resonator, like an organ pipe, that is tuned to a frequency of about 2 cycles per second.

As the level of water in the reservoir drops, less and less energy is available to drive the eruption, and the height of the column decreases. This is the decline phase, which lasts up to about four minutes.

Once the eruption has ceased, there is a period of about 40 minutes when the system is too depleted to sustain another eruption. After that, another eruption is possible; the duration of the quiet period depends on how much water was ejected in the previous eruption. Most eruptions at Old Faithful last about 4 minutes and are followed by a quiet period lasting about 70 minutes. About one-quarter of the eruptions, however, last only about 2 minutes, leaving the reservoir fuller than usual: these eruptions are followed by a quiet period of only about 50 minutes. Why there are these two kinds of eruptions is unknown; the sequence of long and short cycles is not predictable. Some other geysers in the park also have two kinds of eruption. Steamboat Geyser, for example, in the Norris Geyser basin, commonly erupts to height of only 10 or 20 feet, but occasionally it erupts to a height of about 400 feet—three times higher than Old Faithful.

Besides water and steam, the geysers and hot springs bring many dissolved minerals to the surface, some of which precipitate out of solution as the water cools, forming a material called *sinter*. Thus, precipitated silica forms mounds of sinter around the bases of the geysers. At the edges of some of the hot pools, a skin of sinter offers a treacherous illusion of solid ground to the unwary visitor. At Mammoth Hot Springs, in the northern part of the park, the main dissolved mineral is calcium carbonate, which the hot water extracts from limestone strata on its way to the surface. The carbonate is then precipitated out in the form of *travertine* terraces: graceful downward-stepping marble basins that would not look out of place in the gardens of Versailles. Sulfur, iron, and arsenic are among the components of other minerals that form characteristic deposits in or near the hot springs.

The many dissolved minerals, and the high temperatures, foster a zoo of exotic bacteria that are adapted to these environments. The bacteria tinge the pools with characteristic colors: in Porcelain Basin, for example, acid-loving cyanobacteria lend a brilliant green to some of the outflow channels, while other organisms tinge their pools with infinite shadings of red and yellow and brown.

Yellowstone today is a hot-water theme park powered by a friendly, dormant volcano, but it will not stay that way forever. If the Hebgen Lake earthquake altered the timing of Old Faithful's eruptions (and also sent a chimney of Old Faithful Inn tumbling through the roof of the dining hall), we can be sure that much greater changes lie in the future, both for the landscape itself and for humans' use of it.

Among the agents of future change will be both earthquakes and volcanic eruptions. Within the caldera itself, earthquakes will remain frequent but small, as the crust is simply too hot to sustain brittle fractures reaching more than about 5 kilometers below the surface. But on the edges of the Yellowstone Plateau, earthquakes the size of the Hegben Lake earthquake are likely to rock the entire region, further altering the park's hydrothermal systems and endangering the human communities in and around the park.

Volcanic activity may run the entire gamut of Yellowstone's past performances. Because there has not been an actual magmatic eruption over the past 70,000 years, the most likely volcanic events are steam explosions of the kind that gouged out the Pocket Basin crater. The caldera's ongoing cycles of uplift and subsidence, however, mean that it has the capacity to unleash a true magmatic eruption. During the uplift from 1926 to 1985, for example, the caldera's volume increased by about a quarter of a cubic mile. The venting of this amount of magma would correspond roughly to the 1980 eruption at Mount St. Helens. A future eruption could take various forms: a rhyolitic eruption near one of the two resurgent domes or a basalt flow near the old ring fractures that, as at Mammoth, mark the edges of the magma chamber's partially sunken roof.

Eventually—perhaps next century, but more likely hundreds of centuries from now—a new cycle of catastrophic volcanism is likely to begin. When the next caldera-forming eruption takes place, the devastation will extend for hundreds of miles beyond the park's boundaries.

In the very long term, we can be confident that the North American Plate will drift farther to the southwest. Gradually, the Yellowstone Plateau will cool and sink, while locations to the northeast of Yellowstone will feel the heat of the mantle plume for the first time. In 10 million years, Billings, Montana, may well be in the thick of things. But by that time the residents of Billings, along with the rest of the human race, will have evolved into something else and may even be safely ensconced on the planet of some distant star.

Somewhere east of Yellowstone, the terrain levels out. America's midsection—prairie and plain—should be exempt from the geological dramas that built the vertical landscapes of the West. For we are now firmly on the North American Plate, having taken a full thousand miles to cross a plate boundary that is often drawn as a simple line—the San Andreas fault. Yet even here, apparently, we are not entirely safe. Just a few generations ago, the country's heartland was torn by a seismic cataclysm—a cataclysm that, more than any other, illustrates the limitations of our ability to understand or predict earthquakes.

10

A RIVER RUNS THROUGH IT: THE MISSISSIPPI AND THE NEW MADRID EARTHQUAKES

During the winter of 1811–1812, an extraordinary series of earthquakes struck what was then the western frontier of the United States: the Mississippi Valley. The earthquakes were centered in a broad region now encompassed by northeastern Arkansas, southeastern Missouri, western Tennessee, and western Kentucky, but they have been named for the town of New Madrid, which was then a small river settlement 180 miles south of St. Louis in the Missouri Territory. Native

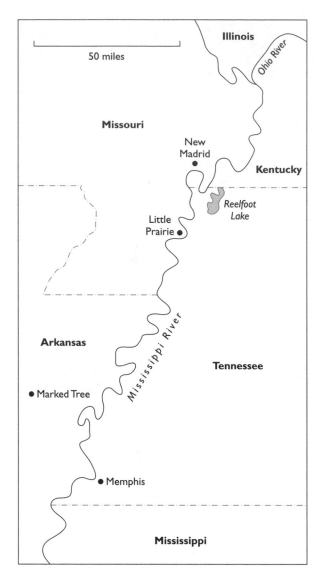

Map of the region of the New Madrid earthquakes of 1811–1812.

Americans, a couple of thousand settlers, riverboatmen, and assorted travelers experienced the violence of the earthquakes at first hand and spread vivid accounts of them to the cities of the East. Yet the earthquakes themselves were speedier messengers: so powerful were some of the shocks that they were felt by residents of almost every city of eastern North America, from St. Louis to Washington, D.C., and from New

Orleans to Quebec City—an area of nearly a million square miles. The seismic energy released during the three-month period of the New Madrid earthquakes, according to estimates by some seismologists, was greater than the total energy released in all other earthquakes in the contiguous United States over the entire course of recorded history.

How could such a powerful series of earthquakes have rocked the stable midsection of the North American Plate, and specifically the placid Mississippi Valley, where no dramatic San Andreas fault gouges the landscape, and no mountain ranges attest to ancient seismic activity? Were those earthquakes something unique and anomalous, or have they occurred repeatedly over geological time? And if the latter, need we prepare for a repeat performance, there or in similar regions? The answers to these questions—still far from certain—have come from studying the accounts of eyewitnesses, monitoring current seismic activity and deformation in the region, and examining the traces that the 1811–1812 earthquake sequence, as well as its predecessors, have left in the ground.

The Earthquakes

The earthquake sequence began with a very powerful shock early on December 16, 1811. One anonymous resident of New Madrid wrote a description of the event that was later published in the *Pennsylvania Gazette:*

> About 2 o'clock this morning we were awakened by a most tremendous noise, while the house danced about and seemed as if it would fall on our heads. I soon conjectured the cause of our troubles, and cried out it was an Earthquake, and for the family to leave the house; which we found very difficult to do, owing to its rolling and jostling about. The shock was soon over, and no injury was sustained, except the loss of the chimney, and the exposure of my family to the cold of the night. At the time of this shock, the heavens were very clear and serene, not a breath of air stirring; but in five minutes it became very dark, and a vapour which seemed to impregnate the atmosphere, had a disagreeable smell, and produced a difficulty of respiration. I knew not how to account for this at the time, but when I saw, in the morning, the situation of my neighbours' houses, all of them more or less injured, I attributed it to the dust. . . . The darkness continued till daybreak; during this time we had eight more shocks, none of them so violent as the first.

At half past 6 o'clock in the morning it cleared up, and believing the danger over I left home, to see what injury my neighbours had sustained. A few minutes after my departure there was another shock, extremely violent. I hurried home as fast as I could, but the agitation of the Earth was so great that it was with much difficulty I kept my balance—the motion of the Earth was about 12 inches to and fro. I cannot give you an accurate description of this moment; the Earth seemed convulsed—the houses shook very much—chimneys falling in every direction. The loud, hoarse roaring which attended the earthquake, together with the cries, screams, and yells of the people, seems still ringing in my ears.

A smaller settlement called Little Prairie, 30 miles downstream from New Madrid and near the bottom of the Missouri boot heel, was affected even more severely by the December 16 earthquake, probably because it was closer to the source. A resident of the settlement, James Fletcher, described what happened:

About 2 o'clock, A.M., we felt a severe concussion of the Earth, which we supposed to be occasioned by a distant earthquake, and did not apprehend much damage. Between that time and day we felt several other slighter shocks; about sunrise another very severe one came on, attended with a perpendicular bouncing that caused the Earth to open in many places—some eight and 10 feet wide, numbers of less width, and of considerable length. . . . Where one of these large openings are, one side remains as high as before the shock and the other is sunk; some more, some less; but the deepest I saw was about 12 feet. The Earth was, in the course of 15 minutes after the shock in the morning, entirely inundated with water. The pressing of the Earth, if the expression be allowable, caused the water to spout out of the pores of the Earth, to the height of eight or 10 feet! The agitation of the Earth was so great that it was with difficulty any could stand on their feet, some could not. The air was very strongly impregnated with a sulphurous smell.

Another account describes the experiences of an elderly miller at Little Prairie:

When the tenth shock occurred, he was standing in his own yard, situated on the bank of the bayou of the Big Lake; the bank gave way, and sunk down about thirty yards from the water's edge, as far as he could see up and down the stream. It upset his mill, and one end of his dwelling house sunk down considerably; the surface on the opposite side of the bayou, which before was swamp, became dry land; the side he was on became lower. His family at this time were running away

from the house towards the woods; a large crack in the ground prevented their retreat into the open field. They had just assembled together when the eleventh shock came on, after which there was not perhaps a square acre of ground unbroken in the neighborhood, and in
about 15 minutes after the shock the water rose round them waist
deep. The old gentleman in leading his family, endeavoring to find
higher land, would sometimes be precipitated headlong into one of
those cracks in the Earth, which were concealed from the eye by the
muddy water through which they were wading. As they proceeded, the
Earth continued to burst open, and mud, water, sand and stone coal,
were thrown up the distance of thirty yards—frequently trees of a large
size were split open, 15 or 20 feet up. After wading eight miles, he
came to dry land.

Little Prairie was completely destroyed. Fletcher and 200 other residents, after camping in makeshift tents for a few days, set out to walk
to New Madrid. Because of the fissures that everywhere obstructed
their progress, it took them three days to cover the distance.

Almost all the accounts of the December 16 mainshock tell of the
loud sound of the earthquake. Many of the aftershocks could also be
heard. The Scottish naturalist John Bradbury, who was on the river
south of Little Prairie when the mainshock struck, described the sound
as that of "the most violent tempest of wind mixed with a sound equal
to the loudest thunder, but more hollow and vibrating." Whereas the
windlike sound might have been the effect of trees being shaken, the
hollow, vibrating noise was certainly produced by the seismic waves
themselves, as they were transmitted out of the ground and were converted from seismic to audible, acoustic waves in the atmosphere.
Luckily, when seismic waves reach the surface, only a tiny portion of
their energy is transferred to the air; most is reflected back into the
ground. Furthermore, humans can hear only high-frequency waves
(above about 20 cycles per second). If Bradbury's ears had been sensitive to frequencies of around 1 cycle per second, he might well have
been permanently deafened!

Sand-blows like those described above were reported by many witnesses at widespread locations. Just as happened during the great subduction earthquakes in the Pacific Northwest (see Chapter 1), sandy,
waterlogged subsoil was liquefied and pressurized by the intense shaking and then ejected through the surface as miniature volcanoes. In
some places, the sand-blows were so numerous that, by the time the
entire series of earthquakes was over, 25 percent or more of the land

Sand-blow deposits (light patches) produced during the New Madrid earthquakes.
(Courtesy of S. Obermeier, USGS)

surface was covered with sand-blow deposits. Even today, nearly 200 years after the earthquakes and after clearing and repeated plowing of the ground, thousands of these sand-blows are readily spotted from the air as round, light-colored patches against darker surrounding soil.

The fissures were caused by uneven sinking of the ground as the water in the subsoil was expelled onto the surface. Along the river, fissures marked where the banks had slumped toward or into the water. Like the sand-blows, many of the fissures are still evident today, even though they have been eroded and filled with soil. Some of them are more than a mile long and many yards wide. The notion that, during earthquakes, the ground opens in yawning chasms that can swallow people whole is usual-ly dismissed as fantasy, but it did happen at New Madrid. One man's expe-rience was described in the *Bairdstown Repository:* "The violent agitation of the ground was such at one time as induced him to hold to a tree to support himself; the Earth gave way at the place, and he with the tree sunk down, and he got wounded in the fall. The fissure was so deep as to put it out of his power to get out at that place. He made his way along the fis-sure until a sloping side offered him an opportunity of crawling out."

Many witnesses described a foul odor that accompanied the larger shocks; most often it was described as "sulfurous." Indeed, the odor was probably a result of hydrogen sulfide and other gases produced by decomposing organic matter in the subsoil—gases that were released by the sand-blows and fissuring.

The December 16 mainshock caused damage far beyond New Madrid. In St. Louis, Cincinnati, and Louisville, chimneys, parapets and gable ends were thrown down. At greater distances, damage was minor, but the mainshock and at least two of the aftershocks were felt by residents up and down the east coast, including President Madison in the White House and ex-president Thomas Jefferson at Monticello. (It was Jefferson, of course, who had arranged the Louisiana Purchase eight years before, thus ensuring that history's most powerful intraplate earthquakes would happen on U.S. soil!)

In more distant cities, the main effects of the earthquake were to ring church bells and to stop domestic pendulum clocks. Both of these effects depend on *resonance:* even a slight, periodic oscillation of the ground, when it matches the natural resonant frequency of a movable object such as a church bell, can gradually bring it into visible motion or, conversely, can gradually bring a swinging object to a standstill. Because church bells were commonly used to rouse the populace in case of fires, many people across America assumed that their cities were in flames.

Aftershocks, some very severe, continued to jolt the region of New Madrid during the ensuing weeks, causing repeated fissuring and sand-blows. Sometimes tremors continued for hours and could be felt as far away as Louisville. One of the aftershocks (probably on January 7) was described by Louis Bringier, an interesting character who had "gone native" after gambling debts forced him to leave New Orleans. He lived among the Indians and is even said to have become an Indian chief, although he later returned to New Orleans and eventually became surveyor-general of the state. He wrote:

> I happened to be passing in [the] neighborhood where the principal
> shock took place. . . . The water that filled the lower cavities . . . rushed
> out in all quarters, bringing with it an enormous quantity of car-
> bonized wood . . . which was ejected to the height of from 10 to
> 15 feet, and fell in a black shower, mixed with the sand which its rapid
> motion had forced along; at the same time, the roaring and whistling
> produced by the impetuosity of the air escaping from its confinement,

seemed to increase the horrible disorder of the trees which everywhere encountered each other, being blown up, cracking and splitting, and falling by thousands at a time. In the meantime, the surface was sinking, and a black liquid was rising up to the belly of my horse, who stood motionless, struck with terror. These occurrences occupied nearly two minutes.

Another person who experienced one of the aftershocks while on horseback was the naturalist John James Audubon, who was traveling in Kentucky. Audubon gave a vivid account of his horse's reaction to the earthquake:

> I had proceeded about a mile, when I heard what I imagined to be the distant rumbling of a violent tornado, on which I spurred my steed, with a wish to gallop as fast as possible to a place of shelter; but it would not do, the animal knew better than I what was forthcoming, and instead of going faster, so nearly stopped that I remarked [noticed] he placed one foot after another on the ground, with as much precaution as if walking on a smooth sheet of ice. I thought he had suddenly foundered [gone lame], and speaking to him, was on the point of dismounting and leading him, when he all of a sudden fell a-groaning piteously, hung his head, spread out his four legs, as if to save himself from falling, and stood stock still, continuing to groan. I thought my horse was about to die, and would have sprung from his back had a minute more elapsed, but at that instant all the shrubs and trees began to move from their very roots, the ground rose and fell in successive furrows, like the ruffled waters of a lake.

Audubon's horse evidently sensed the feeble, early arriving primary and secondary seismic waves, followed seconds later by higher amplitude surface waves. But the origin of the waves seen by Audubon (and by many other people during large earthquakes) remains a complete mystery. To be visible as waves moving across the surface, they would have to be of shorter wavelength and slower velocity than anything that has been recorded by seismographs. Some seismologists have suggested that they are akin to sea waves and are made possible by intense liquefaction of the ground. But Audubon's account does not suggest that liquefaction occurred in his vicinity. Other seismologists, despairing of a physical explanation, have proposed that the visible waves are some kind of optical illusion caused by vibration of the viewer's eyes or head.

At nine in the morning of January 23, a second mainshock, almost as powerful as the first, struck the region. There were no boats on the

Mississippi at the time—traffic had been interrupted by the icing-over of the Ohio River—and accounts from New Madrid are sparse. One resident described the January 23 earthquake as being "as violent as the severest of the former ones" and said that it was followed by a stream of aftershocks that kept the ground in continuous visible motion for 12 days.

Then, at 3.45 A.M. on February 7, came the coup de grace: a third mainshock, as powerful as the first. The epicenter of this shock was evidently much nearer to New Madrid than the earlier shocks, because the destruction there was total: buildings that earlier had merely lost their chimneys were now leveled, or nearly so. Luckily, the inhabitants had long since moved out of their houses into temporary encampments, so there were no fatalities. In addition, the 25-foot-high riverbank on which the town was built sank by 15 feet. This exposed the town to the action of the spring floods, which did indeed wash away the remains of the town a few weeks later. The Mississippi, which even before the earthquake was encroaching on the town, today flows over the place where it stood, and a "new" New Madrid has risen, a mile or more north of its ill-fated predecessor.

The most dramatic effects of the February mainshock, however, were on the river, whose bed was lifted broadly up in the vicinity of New Madrid, briefly halting or even reversing the river's flow, and creating waterfalls or rapids at two locations where none had been before. Because the ice on the Ohio had been loosened by the January 23 mainshock, many boats were on the river, and some were destroyed, probably with loss of life. One survivor, Matthias Speed, wrote of his experiences:

> In descending the Mississippi, on the night of the 6th February, we tied our boat to a willow bar on the west bank of the river, opposite the 9th island counting from the mouth of the Ohio. We were lashed to another boat. . . . About 3 o'clock, on the morning of the 7th, we were waked by the violent agitation of the boat, attended with a noise more tremendous and terrific than I can describe or any one can conceive, who was not present or near to such a scene. The constant discharge of heavy cannon might give some idea of the noise for loudness, but this was infinitely more terrible, on account of its appearing to be subterraneous.
>
> As soon as we waked we discovered that the bar to which we were tied was sinking, we cut loose and moved our boats for the middle of the river. After getting out so far as to be out of danger from the trees

which were falling in from the bank—the swell in the river was so
great as to threaten the sinking of the boat every moment.

At day light we perceived the head of the tenth island. During all
this time we had made only four miles down the river—from which
circumstance, and from that of an immense quantity of water rushing
into the river from the woods—it is evident that the Earth at this place,
or below, had been raised so high as to stop the progress of the river,
and caused it to overflow its banks. We took the right hand channel of
the river, and having reached within about half a mile of the lower end
of the island, we were affrightened with the appearance of a dreadful
rapid of falls in the river just below us; we were so far in the suck that
it was impossible now to land—all hope of surviving was now lost and
certain destruction appeared to await us! [Yet we] passed the rapids
without injury, keeping our bow foremost, both boats being still lashed
together.

As we passed the point on the left hand below the island, the bank
and trees were rapidly falling in. From the state of alarm I was in at this
time, I cannot pretend to be correct as to the length or height of the
falls; but my impression is, they were about equal to the rapids of the
Ohio [which descended 23 feet over 2 miles]. As we passed the lower
point of the island, looking back, up the left channel, we thought the
falls extended higher up the river on that side than the other.

The water of the river, after it was fairly light, appeared to be
almost black, with something like the dust of stone coal—We landed at
New Madrid about breakfast time, without having experienced any
injury. The appearance of the town, and the situation of the inhabitants,
were such as to afford but little relief to our minds. . . . There was
scarcely a house left entire—some wholly prostrated, others unroofed
and not a chimney standing.

Two sets of rapids appeared at the time of the earthquake: one
(crossed by Matthias Speed) was just above New Madrid; the other,
equally high, was about 8 miles downstream. As we will discuss later, the
appearance of the rapids is a clue to the underlying mechanism of the
earthquake. Apparently, the disturbance of the riverbed caused the river
to run backward for a while; a resident of New Madrid named Eliza
Bryan described a "retrograde current, rapid as a torrent." But the river
soon resumed its normal course, and the rapids were smoothed out by
the flow of the river within a few days. There were other more permanent signs of the changes in ground level: the Reelfoot River, a small
tributary that joins the Mississippi near New Madrid, was blocked by
uplift of a portion of its bed, causing an 18-mile-long body of water to

collect on the sunken land upstream and to the northeast. Reelfoot Lake persists to this day.

So great was the devastation caused by the third earthquake, and so unremitting the aftershocks, that the inhabitants of New Madrid and all the nearby settlements fled the scene in despair. The refugees became a common sight in surrounding towns. "No pencil can paint the distress of the many movers!" wrote a correspondent in Russelville, Kentucky. "Men, women and children, barefooted and naked! Without money and without food."

Although damage in the nearest cities, such as St. Louis and Louisville, was not extreme, the shocks and aftershocks, which persisted for months, put a significant psychological strain on the inhabitants. Some dealt with this strain by developing an almost obsessive interest in the minutiae of the earthquakes. Jared Brooks, an engineer in Louisville, set up a system of pendulums and spring-mounted weights so that he could record the strength, duration, and direction of every shock, along with details of the weather on each occasion (supposing that the weather could somehow help explain the shocks' occurrence). By March 15, he had made annotations on no less than 1874 shocks— and this at a distance of more than 200 miles from the earthquakes' epicenters. Others felt the same need to memorize the shocks, but lacked the skill to do so. One judge living in Michigan wrote, "We have had nine shocks of the earthquake here, of which I have an exact memory of eight and have somehow entirely lost the time of the other. I felt four myself. I know only one person, a French lady, who felt the whole; speaking here of the eight."

Naturally, the cause of the earthquakes was much discussed. Considering the time and place, it is remarkable how committed some people were to the principle of physical causation, even if the causes suggested were far off base. A subscriber to the Louisiana *Gazette* of St. Louis wrote:

> According to the hypothesis of some, earthquakes are occasioned by subterranean fires throwing down the arches or vaults of the Earth; according to others the rarefaction of the abyss waters, interior combustion and fermentation, volcanic operations, and lately by the electric fluid. The latter hypothesis seems to be most accredited, as it is evidently the most rational. The instantaneous effects of the earthquakes prove beyond doubt that electricity is the principal agent in this alarming and terrible phenomenon.

Others hedged physical explanations with more traditional explanations of the fire-and-brimstone variety. A correspondent to the *Louisiana Gazette and Daily Advertiser* of New Orleans wrote:

> The comet has been passing to the westward since it passed its perihelion—perhaps it has touched the mountain of California, that has given a small shake to this side of the globe—or the shake which the Natchezians have felt may be a mysterious visitation from the Author of all nature, on them for their sins—wickedness and want of good faith having long prevailed in that territory. Sodom and Gomorra would have been saved had three righteous persons been found in it—we therefore hope that Natchez has been saved on the same principle.

Dramatic as the earthquakes were, public interest in them was soon extinguished by the press of more exciting events—the early battles of the War of 1812, and Napoleon's disastrous invasion of Russia. What is more, the accounts of the devastation, coming as they did from the Wild West, were met with a dampening skepticism in the East. Even geologists placed little faith in the reports. As the years went by, it seemed as if less and less had happened—perhaps nothing at all. In 1883, a scientist named James MacFarlane presented a paper at the American Association for the Advancement of Science titled "The 'Earthquake' at New Madrid, Missouri, in 1811, Probably Not an Earthquake." Memories of the earthquakes, it seemed, had been erased as thoroughly as New Madrid itself.

After the 1906 San Francisco earthquake, however, there was renewed scientific interest in America's seismic history. In particular, USGS geologist Myron Fuller began to gather together the scattered contemporary accounts, both from the New Madrid area and from distant cities, with the aim of establishing the earthquakes' magnitude. This interest was rekindled in the 1970s by Otto Nuttli of St. Louis University. (Most of the accounts cited above come from these sources.) According to the most current assessment, by Arch Johnston and Eugene "Buddy" Schweig of the University of Memphis and the USGS, the three mainshocks had magnitudes of about 8.1, 7.8, and 8.0. Even some of the aftershocks were of enormous magnitude: an aftershock that struck six hours after the December 16 mainshock had an estimated magnitude of 7.2—much bigger than the mainshock of the 1994 Northridge earthquake. Clearly, the New Madrid earthquakes were extraordinary events.

The Underground Landscape

"Flat as a pancake" would be the average visitor's description of the Mississippi River floodplain. But the practiced geologist's eye can pick out subtle variations on the theme of horizontality—slight rises, scarcely high enough to block one's view; gentle hollows where water accumulates, nourishing the bald cypress trees; little slopes that take a dozen strides to climb or descend. The river too senses the topography: where the states of Kentucky, Missouri, and Tennessee come together, the river takes a 15-mile northward loop—the Kentucky Bend—just to avoid an area of elevated ground known as the Tiptonville Dome. The "new" New Madrid sits at the northern limit of the loop.

As we'll discover shortly, some of these subtle features turn out to offer important clues to what happened in the winter of 1811–1812.

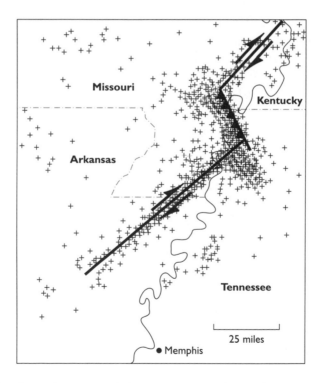

Recent earthquakes in the New Madrid region appear to define three fault zones responsible for the three great 1811–1812 earthquakes. Two of these faults involve right-lateral slip, and one involves reverse slip (toothed line). (Adapted from an illustration by D. Russ, USGS)

But more direct evidence comes from the many small earthquakes that continue to strike the area today. The epicenters of these small earthquakes, which are picked up by the local seismographic network, lie along three linear segments that together form a zigzag line. The longest of the three, called the Blytheville seismic zone, runs northeastward from the vicinity of Marked Tree, in eastern Arkansas, across the southeast corner of the Missouri boot heel, and into northern Tennessee, a distance of about 70 miles. Analysis of the waveforms of the earthquakes along this segment indicate that they are caused by right-lateral strike-slip ruptures on fault planes oriented along the line of seismicity. In other words, when an earthquake occurs, rock on the northwest side slips northeastward relative to rock on the southeast side. A second, much shorter segment, also characterized by right-lateral movement, extends from the boot-heel region a short distance farther northeastward into eastern Missouri. This segment, called the New Madrid north zone, looks like a continuation of the first segment, except that it lies about 25 miles to the northwest. Connecting these two segments is another, broader region of seismicity that crosses the Mississippi in the neighborhood of New Madrid. The small earthquakes in this region have "reverse" mechanisms: rock on the southwest side moves upward and northeastward relative to rock on the northeast side. The overall pattern of seismicity suggests that this general region of the Earth's crust is being subjected to east-west shortening and north-south stretching.

It seems likely that the three segments outlined by current seismic activity are faults that ruptured in the 1811–1812 earthquake sequence. Because the sequence involved three mainshocks, it is tempting to assume that each was caused by rupture of one of the three segments. But before we accept this assumption, we need to take a look at the ground surface overlying the segments.

The ground over the middle, "stepover" segment tells a clear story. A northeast-facing scarp named the Reelfoot scarp, about 6 to 10 feet high, runs for 20 miles along the zone of seismicity. It runs along the western edge of Reelfoot Lake, on the east side of the river in Tennessee. Farther to the northwest, it runs diagonally across the piece of land contained within the Kentucky Bend, and then it reappears on the west side of the river, at the western edge of the modern town of New Madrid.

Seismic reflection studies have revealed the presence of a fault dipping downward to the southwest from the scarp. Thus, this fault—the

13 Disaster scene in Granada Hills five hours after the Northridge earthquake. Homes were destroyed by fire caused by a broken gas main. (Kerry Sieh and Simon LeVay)

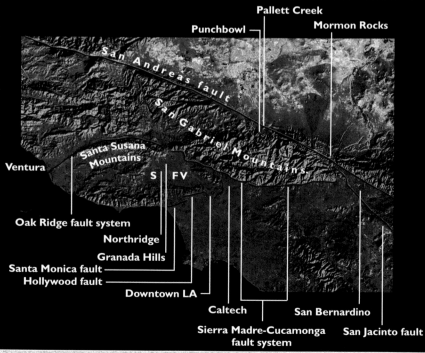

14 Southern California faults and localities mentioned in the text. The background is a combined shaded-relief map and satellite image. (Adapted from image provided by R. Crippen, JPL)

15 View northward from the star-studded intersection of Hollywood and Vine. The ramplike rise in the street just beyond the Capitol Records building is the scarp of the Hollywood fault. Beyond the scarp and the freeway bridge are the Hollywood Hills, raised by innumerable ruptures of the fault (Kerry Sieh and Simon LeVay)

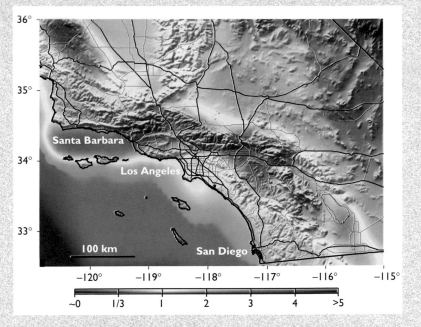

16 Map of Southern California, showing the number of occasions that each point is predicted to experience seismic shaking of 0.2 g or greater within the next 100 years. (Courtesy of K. Hudnut and the Southern California Earthquake Center)

17 Tree die-off caused by carbon dioxide seepage on the shore of Horseshoe Lake, southeast flank of Mammoth Mountain. (Kerry Sieh and Simon LeVay)

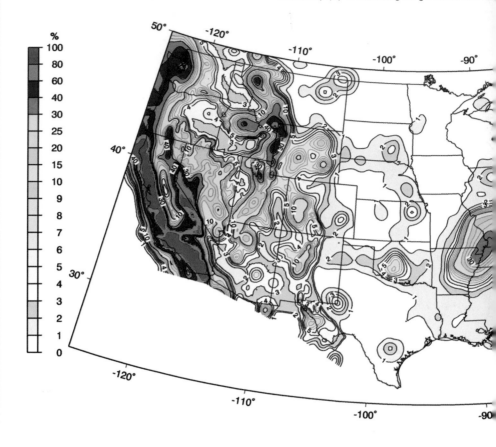

18 Map of seismic hazard in the "lower 48" states. Colors represent the probability that shaking of 0.2 g or greater will occur within the next 200 years. (Courtesy of A. Frankel, USGS)

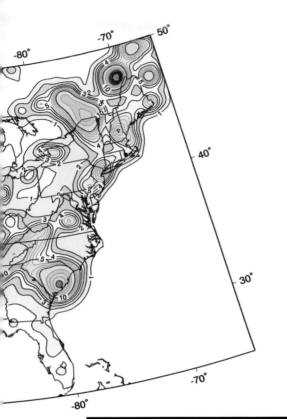

19 Calderas of the Snake River Plain and Yellowstone region, showing the ages of calderas (in millions of years) and current seismicity. The progression of ages indicates the movement of the North American Plate over the Yellowstone hot spot during the past 16 million years.
(Based on a compilation of data by R. Smith, University of Utah)

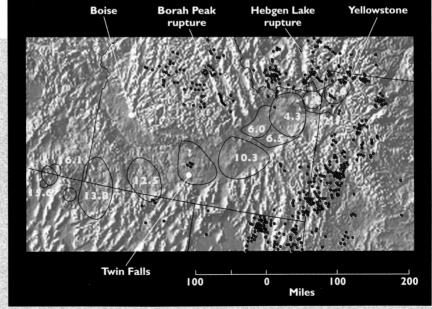

Boise   Borah Peak rupture   Hebgen Lake rupture   Yellowstone

4.3
6.0
6.5
10.3
16.1
15.0   13.8   12.5

Twin Falls

100   0   100   200
Miles

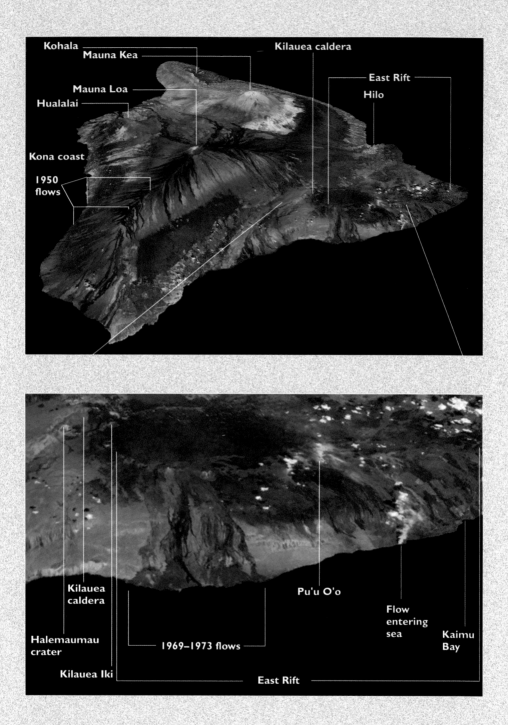

20　The Big Island of Hawaii seen from the southeast in a simulated perspective view in which the vertical dimension is exaggerated twofold. Lava flows are in blue and green, vegetation in red. To create the image, Oliver Chadwick and Steven Adams, JPL, superimposed SPOT multispectral satellite data (shown by permission) on a digital elevation model.

21 An a'a lava flow on Kilauea. (USGS)

22 Solidified pahoehoe lava from the south-west rift zone of Kilauea. (Kerry Sieh and Simon LeVay)

23 Aerial photographs of the Kalapana and Kaimu Bay area, showing its progressive destruction by lava flows from Kilauea's east rift zone. In the top frame (May 2, 1990) the lava has crossed Highway 130 (upper left) and destroyed much of the Kalapana subdivision. In the middle frame (August 20, 1990) the remainder of Kalapana has been overrun and the lava is entering the ocean. In the bottom frame (December 7, 1990) the lava has filled Kaimu Bay. (USGS)

Reelfoot fault—has all the signs of being a reverse fault, whose rupture in 1811–1812 raised the Reelfoot scarp, or some part of it.

Neither of the two segments of right-lateral seismicity—the Blytheville and New Madrid north zones—have obvious traces along the surface. If, as is generally assumed, these two segments are indeed active strike-slip faults that ruptured in 1811–1812, the rupture planes were probably deep in the basement rocks and did not propagate through the overlying sedimentary layers to the surface. Nevertheless, there may well have been deformations of the ground surface related to the deep ruptures. For example, it appears that two sunklands (broad, swampy depressions) on the northwest side of the Blytheville seismic zone, known as Big Lake and Lake St. Francis, either formed or were deepened during the 1811–1812 earthquakes.

It seems most likely that the first earthquake of the sequence, the December 16 mainshock, was caused by right-lateral slip on the fault underlying the Blytheville seismic zone. The main evidence for this belief is the reports of shaking intensity. Locations such as Little Prairie, near the Blytheville fault zone, were much more severely shaken in the December 16 mainshock than were locations to the northeast, such as New Madrid. Also, it is appropriate to assign the most powerful earthquake of the entire sequence to what is probably the longest fault in the area. Some experts, however, have suggested that other faults in the area may have broken in the same mainshock or in one of the powerful aftershocks.

Johnston and Schweig attribute the January 23 mainshock to a rupture along the shorter, northern zone of right-lateral seismicity (sometimes called the New Madrid north zone) or along other poorly defined zones of seismicity in that general neighborhood. Their main reasons for making this attribution are threefold. First, the January mainshock produced very intense shaking at New Madrid. Second, the January mainshock had a distinctly smaller magnitude that the other two mainshocks, as judged by the reports from distant cities; this fits best with rupture along the relatively short New Madrid north seismicity zone. Third, there are no accounts of disruption of the Mississippi riverbed in connection with the January mainshock; therefore, the Reelfoot fault, which crosses the river, is not a good candidate for the location of the January rupture.

There is general agreement that the third mainshock, on February 7, was caused by rupture of the Reelfoot fault. This fault crosses the river

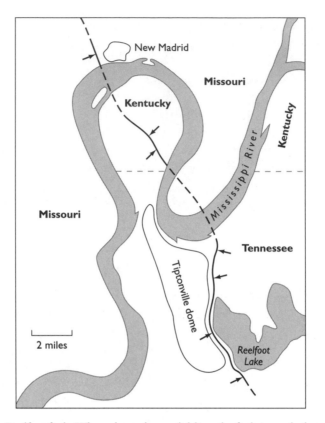

Map of the Reelfoot fault. Where shown by a solid line, the fault is marked at the surface by a visible scarp. Where shown by a dashed line, the scarp has been washed away by the river. The arrows mark the locations where the fault has been excavated. Sudden dropping of the land northeast of the fault during the February 1812 earthquake resulted in dramatic changes in the flow of the Mississippi River and in the formation of Reelfoot Lake. (Adapted from an illustration by R. Van Arsdale and others)

three times, making it the natural candidate for the source of an earthquake that was accompanied by major disruption of the riverbed. Furthermore, previous ruptures of the Blytheville and New Madrid North zones would have transferred stresses onto the Reelfoot fault, bringing it closer to failure.

Trying to fit the topography of the Reelfoot fault to the details of the historical accounts, however, has proved a little frustrating. The accounts, such as that of Matthias Speed quoted earlier, seem to call for four separate disruptions of the riverbed: a barrier to the river's flow somewhere upstream of Island Number 10; a waterfall or rapids right at Island Number 10; a second flow barrier just downstream of New Madrid (which produced the retrograde current described by Eliza Bryan); and a second waterfall or rapids about 8 miles downstream of New Madrid.

The Mississippi River no longer runs precisely in the same channel as it did in 1812, nor does Island Number 10 exist today. According to a contemporary map, however, Island Number 10 was located at the beginning of the Kentucky Bend, right where the Reelfoot fault first crosses the river. But reverse-fault slip would raise the southwest side of the fault, producing a barrier to the river's flow at this location, not a waterfall as described by Speed. To get around this problem, one can hazard the guess that the island was actually located farther around the Kentucky Bend, at the place where the fault crosses the river for a second time. Since the river at this location is heading northward, reverse-fault slip would produce a waterfall. The upstream flow barrier described by Speed can then be attributed to reverse faulting where the fault first crosses the river.

This leaves the second flow barrier, just south of New Madrid, and the second waterfall, 8 miles downstream. The New Madrid flow barrier is neatly accounted for by the third crossing of the river by the Reelfoot fault. The downstream waterfall, on the other hand, remains unaccounted for. Keith Kelson, one of the scientists who has studied the fault, suggests that Speed meant to say "upstream" of New Madrid; in other words, that this was the same waterfall that he himself went over. But his account isn't easily compatible with a verbal slip of that kind. Roy Van Arsdale, a geologist at the University of Memphis who has carried out many field studies of the New Madrid earthquakes, believes Speed's account and intends to go looking for signs of a rupture at the indicated downstream location. He suggests that the waterfall was produced by upward movement on a "backthrust" fault that splays upward from the Reelfoot fault. In other words, a large chunk of the hanging wall of the Reelfoot fault, bounded to the northeast by that fault and to the southwest by the backthrust fault, was wedged upward during the earthquake.

If we assume that the geology and the historical accounts can be shoehorned into reasonable correspondence, there still remains the question of the magnitude of the February 7 mainshock. A 25-mile fault (the extent of the Reelfoot fault that has been mapped) would not generally be thought capable of sustaining an earthquake of moment magnitude 8, the magnitude estimate by Johnston and Schweig. This suggests one of three possibilities: that Johnston and Schweig's estimate is too high, that another fault broke in addition to the Reelfoot fault, or that the Reelfoot fault is longer than the portion that has been mapped. Roy Van Arsdale argues for the last of these three possibilities; he suggests that the fault extends under the bluffs that line the eastern edge of the Mississippi Valley but is not easily detected there because of the irregular topography.

Unearthing Earthquake History

Field geologists have combed the New Madrid area looking for clues to the 1811–1812 earthquakes and to possible earlier earthquakes in the same seismic zone. One interesting set of observations comes from tree ring studies carried out by Roy Van Arsdale in collaboration with dendrochronologists from the University of Arkansas. These scientists took cores from bald cypress trees in Reelfoot Lake, the shallow body of water immediately to the northeast of the Reelfoot scarp. They found that many of the trees showed a major growth spurt during the 3- to 5-year period after 1812. During this brief period, the trees grew as much as five times faster than they did at any other time during their lifetimes. Apparently, then, the New Madrid earthquakes promoted the trees' growth—the exact opposite of the growth-retarding effect of earthquakes on the San Andreas fault, as mentioned in Chapter 4. Presumably the growth-promoting effect was a result of slip on the Reelfoot fault. By rising during the earthquake, the scarp blocked what had been an inconsequential stream, the Reelfoot River, and turned it into a broad, shallow lake, an environment in which bald cypress trees thrive. Thus, the tree ring data are consistent with the belief that the Reelfoot fault is a southwest-dipping reverse fault that slipped in 1811–1812.

To gain some idea of the hazard posed by the New Madrid seismic zone today, it is obviously important to discover whether events similar to those that occurred in 1811–1812 have happened repeatedly over time, and if so, how often. We lack the kind of information that is available about the San Andreas fault and that would allow us to estimate the recurrence interval between great earthquakes: that is, we do not know the long-term slip rate on the relevant faults, nor do we know how far the faults slip during each earthquake. But there is some evidence of earthquakes preceding the 1811–1812 events.

A British geologist, Charles Lyell, who toured the New Madrid area after the earthquakes, reported that the local American Indians recounted a legend of a great earthquake that had devastated the same region at some time in the past. Although Lyell was skeptical of the accounts, they have since been verified by paleoseismological investigations.

Sand-blows and other liquefaction features have been particularly useful in these investigations. Of the innumerable sand-blow deposits now visible in the New Madrid area, most date from the 1811–1812 earthquakes; but some are relics of much earlier earthquakes in the same area. We know this because Native Americans took advantage of some of the sand-blow pits for cooking and other purposes, leaving fragments of pottery, charcoal, and other material inside them. Martitia Tuttle, of the University of Maryland, College Park, as well as a number of colleagues, was able to date some of the potsherds by their style of manufacture—a classic dating technique in archeology. They also dated the charcoal by the radiocarbon method. By these means, Tuttle and her colleagues concluded that an earthquake or sequence of earthquakes strong enough to cause liquefaction struck the New Madrid seismic zone sometime between 1180 and 1400. They also found buried sand-blows that dated to even earlier times—one of them formed four to five thousand years ago.

More evidence of ancient earthquakes has come from excavations on the Reelfoot scarp done by David Russ, Keith Kelson, Roy Van Arsdale, and others. The trenches have not revealed a major fault that reaches the surface at the Reelfoot scarp. Nevertheless, the scarp is in line with the fault visualized underground by the seismic reflection technique. Therefore, the scarp is the surface manifestation of a fold caused by the rupture at depth. Presumably the fracture propagated to some point near the surface, but above that point the slip was accommodated by warping of the sandy, silty, and clayey layers deposited by the Mississippi River.

Although the main fault trace is not evident in the trenches, there are other, secondary features that have cast important light on the seismic history of the New Madrid region. In particular, Kelson and colleagues found a crack in the ground, running along the top of the scarp. The crack resulted from folding of the soil layers into a tighter curve than they could withstand. The crack was filled with clay and other material washed in from neighboring parts of the scarp—probably during or after the 1812 earthquake. But below this material Kelson found soil layers that had subsided into the crack at earlier times, presumably in association with earthquakes preceding the 1812 earthquake. By carbon dating wood specimens found in these layers, Kelson was able to come up with approximate dates for these earlier events. One occurred between about 780 and 1000 and the other between about 1260 and 1650. This latter age range agrees well with the age range deduced by Tuttle from the sand-blow excavations. When one adds the 1812 event, Kelson's dates suggest that the last few ruptures of the Reelfoot fault have taken place at intervals of about 400–500 years.

When faults rupture repeatedly, there must be some ongoing process of crustal deformation that is causing the ruptures. Just as the shearing motion between the Pacific and North American plates, which is responsible for ruptures of the San Andreas fault, can be detected by GPS and other space-based techniques, so the motion that gives rise to ruptures in the New Madrid zone should also be detectable, in principle at least. Because the rate of deformation is likely to be quite a bit slower than the deformation that takes place across the San Andreas fault, it might be difficult to detect it by performing GPS measurements spaced, say, five years apart. But in the early 1990s, a group of seismologists from Stanford University (Lanbo Liu, Mark Zoback, and Paul Segall) used the GPS technique to resurvey a network of markers that had originally been surveyed, by the traditional methods, in the 1950s. With the benefit of a 40-year interval between the surveys, it was a relatively easy matter to detect an ongoing process of deformation. The crust in the area of the New Madrid seismic zone, according to their measurements, is being sheared in a right-lateral sense at a rate of 5 to 7 millimeters (0.2 to 0.3 inch) per year. The findings of a "hybrid" study are not as conclusive as when both sets of measurements are done with the same technique, however, and ongoing GPS measurements have so far failed to verify the Stanford group's report.

This issue leads to a significant question: Have the faults of the New Madrid seismic zone been slipping repeatedly over long periods of geological time? There is very little evidence aboveground supporting such a scenario. Certainly, there has been some uplift of the ground to the southwest of the fault, presumably related to a series of earthquakes over many centuries, and this uplift is probably responsible for the gradual

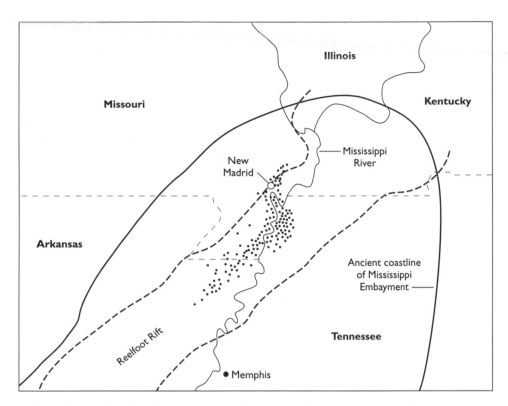

The Reelfoot Rift, within the Mississippi Embayment, and current seismicity in the New Madrid area (dots). (Adapted from an illustration by T. Hildenbrand and J. Hendricks, USGS)

diversion of the Mississippi northward around the Kentucky Bend. But the Reelfoot fault has not raised a mountain range, like those raised along reverse faults in the Los Angeles basin.

Evidence from below ground also suggests that long-term slip has been very modest. The seismic reflection studies of the Reelfoot fault are particularly informative: where 40-million-year-old sedimentary layers cross the fault, they are offset vertically by no more than about 50 feet. The two strike-slip seismicity zones don't seem to have slipped much either; according to magnetic studies, buried rock formations that are more than 300 million years old have been offset horizontally by no more than 6 miles. By comparison, the San Andreas fault slips by this distance in a mere 300 *thousand* years!

So what is going on? Is the New Madrid seismic zone a geological youngster—perhaps no more than a few tens of thousands of years old? Or is it a much older system whose long-term slip rate is far less than its recent seismic history would suggest?

A hint of an answer comes from excavations on a fault 500 miles to the west of New Madrid, in southern Oklahoma. Anthony Crone (of the USGS at Golden, Colorado), Kenneth Luza (of the Oklahoma Geological Survey), and Keith Kelson have cut trenches across this structure, which is named the Meers fault. They found evidence of a major rupture that happened about 1200 years ago and of another that happened about 2800 years ago. But the rupture before that happened more than 100,000 years ago. In other words, earthquakes on the Meers fault may cluster in time, with extremely long intervals between the clusters. If there is a similar temporal clustering in the New Madrid zone, this could help explain why the overall slip is much less than would be expected on the basis of the region's recent seismic history and deformation rates.

The Failed Rift

Despite all that has been discovered about the New Madrid seismic zone, there still remains a central mystery: What is special about this region that causes it to be so plagued by great earthquakes? It is, after all, deep within a tectonic plate and should be relatively protected from the stresses that arise as plates collide, grind past each other, or tear apart. Although there is not yet a definitive answer to this question, some geologists (especially the University of Memphis group) believe that a structure deep beneath the Mississippi Valley may play a role in causing the ongoing deformation and seismicity.

One way to peer beneath the surface is by mapping slight variations in the Earth's magnetic field, which are induced by variations in the magnetic properties of subterranean rocks. Such studies, carried out by USGS scientist Thomas Hildenbrand in the early 1980s, revealed a major anomaly in the granite basement underlying the sedimentary layers. Throughout much of the region, the basement lies about 6000 feet below the surface. But along a 40-mile-wide strip running northeastward across eastern Arkansas, across the Missouri boot heel, and into western Kentucky, the basement has sunk to twice that depth.

The Reelfoot Rift, as this sunken valley has been named, was produced by stretching or extension of the Earth's crust. The walls of the valley mark where normal faulting allowed the central strip to collapse downward as the regions to the northwest and southeast pulled apart from each other.

The rift formed about 600 million years ago, when roachlike trilobites ruled the Earth's shallow seas. It was one arm of a star-shaped array of rifts that marked where a megacontinent was about to break apart. Other arms, to the south and east, eventually opened up to form the Iapetus Ocean, predecessor to our Atlantic Ocean, but the Reelfoot Rift, and another arm extending under what is now southern Oklahoma, were left as "failed rifts."

During the mid-Cretaceous period, about 100 million years ago, the Reelfoot Rift was the scene of renewed activity. Magma rising from the upper mantle forced its way up along the old normal faults, crystallizing into masses of igneous rock—plutons—at scattered locations along the rift. Randel Cox and Roy Van Arsdale have suggested that the source of heat for this episode was a hot spot. Building on earlier suggestions by W. Jason Morgan, Cox and Van Arsdale proposed that a hot spot that now lies near Bermuda was then located under the Reelfoot Rift.

In Cox and Van Arsdale's model, not only did the Bermuda hot spot cause intrusions of magma, it also heated the entire floor of the rift, which caused it to dome upward, reversing the original downward slippage on the normal faults. This upward motion lifted the ground surface into a range of hills, but in time the hills eroded away. Thus, when the region moved off the hot spot and the crust cooled and sank, a troughlike depression was left. This depression became a northward arm of the Gulf of Mexico, the Mississippi Embayment. The embayment gradually filled with marine sediment. Then, at the end of the Cretaceous period, the oceans retreated, and the Mississippi Embayment became the floodplain of the Mississippi River.

The main evidence that the Reelfoot Rift is somehow connected with the New Madrid earthquakes is simply the observation that the Blytheville seismicity zone runs along the center of the rift. This could be a coincidence, but it could also mean that the existence of the rift leads in some manner to the concentration of tectonic stresses, which in turn predisposes it to faulting and earthquakes. Somewhat strengthening this view is the fact that the other failed rift in the region—the one in southern Oklahoma—underlies the Meers fault, whose seismic history has already been described. However, a detailed theory linking ancient rift structures and current earthquakes is still lacking.

A City in Peril

Memphis, Tennessee, did not exist in 1811, but we know that the "fourth Chickasaw Bluff," where the city was later built, was very severely shaken during the December 16 earthquake. It is now a city of over 600,000 people. Several factors—the proximity to the Blytheville seismicity zone; the weak, waterlogged ground on which the bluffs stand; and the general failure of developers and local officials to consider seismic hazards during the construction of the city—make Memphis a candidate for a large-scale seismic disaster.

There are no data to suggest that Memphis, or the New Madrid area in general, is overdue for a repeat of the great earthquake sequence of 1811–1812. If the approximately 400-year recurrence interval suggested by the trenching studies is correct, then the time that has elapsed since 1811–1812 is only about half the average recurrence interval. Therefore, it may well be another two or three hundred years before the events of 1811–1812 repeat themselves. But even without great earthquakes, the region is endangered by moderate ones. Between 1812 and the present, the New Madrid area has been struck by at least 28 damaging earthquakes. In 1843, for example, an earthquake with an estimated magnitude of 6.4 occurred near Marked Tree, at the southern end of the Blytheville seismicity zone, causing severe damage at the present site of Memphis. An even stronger earthquake, about magnitude 6.8, occurred near Charleston, Missouri, in 1895: it was strong enough to cause ground liquefaction over a small area, and it was felt from Ontario, Canada, to New Orleans.

On national seismic hazard maps developed by the USGS, Memphis sits near a bull's-eye of increased hazard associated with the New Madrid area (see Color Plate 18). The official estimate is that Memphis has a 15–20 percent chance of experiencing ground acceleration of 0.2 g or more within a period of 100 years. The USGS maps show the hazard level in the New Madrid zone to be higher than at any other location in the United States east of the Rockies. Furthermore, the USGS predictions relate to firm-rock sites, which are more or less absent in the New Madrid hazard zone. Sites consisting of soft soil, as we've noted in previous chapters, shake much more severely for a given magnitude of earthquake. Therefore the actual chances of experiencing a 0.2 g acceleration in the next 100 years are probably higher, at most sites, than the 15–20 percent estimated by the USGS.

The residents of New Madrid itself are well aware, perhaps even proud, of their town's extraordinary seismic history. Disastrous earthquakes (and equally disastrous floods) are the sleepy community's major claim to fame. If the 1811–1812 sequence does repeat itself and the town is leveled, a third New Madrid will doubtless be constructed, perhaps another mile to the north. The second New Madrid will be ceded to the river, and the first New Madrid, perhaps, will reemerge on the Kentucky shore.

The events at New Madrid are by no means unique: several apparently stable continental regions have been struck by sequences of large earthquakes. Australia offers one dramatic example. The Australian continent is an island in the middle of the Australian Plate: no plate boundary comes anywhere near the continent itself. The interior of the continent is a flat and ancient landmass with little evidence of recent tectonic activity, and until 1968 significant earthquakes were unknown there. Then, in a period of 20 years, Australia was hit by five surface-rupturing earthquakes. The first three occurred northeast of Perth, in Western Australia, but the most recent two happened right in the center of the country, not far from Alice Springs. The second of these—the Tennant Creek earthquake of January 22, 1988—was actually a series of three shocks, each with a magnitude of around 6.5, that took place within the space of 12 hours. Each was caused by rupture of a different fault. According to Anthony Crone, who studied the scarps produced during these shocks, the three faults last ruptured more than 50,000 years ago. Here again, we see an example of earthquakes that cluster in time and space.

Another midplate earthquake struck the flat interior of the Indian continent in September 1993. Named the Latur earthquake, it had a magnitude of 6.4, less powerful than the earthquake that struck Northridge, California, a few months later. But it killed 10,000 people.

Within the interior of the North American Plate, the New Madrid area is not the only place susceptible to earthquakes. From the Mississippi Valley we travel to the eastern seaboard of the United States, a region where, over the course of the past hundreds of millions of years, continents have repeatedly collided and split apart. The last such splitting took place 180 million years ago, and so one would not expect earthquakes there today. Indeed, the East Coast is relatively quiet seismically. But occasional destructive earthquakes continue to occur there, and these earthquakes are made all the more damaging by the general unpreparedness of the population.

11

ALONG THE EASTERN SEA: EARTHQUAKES ON THE ATLANTIC COAST

R*espicit terram et tremit*—"He looks over the Earth, and it trembles." This quotation from Psalm 104 is engraved over the entrance to the Weston Observatory, a modest red-brick building tucked away among the leafy, affluent suburbs west of Boston. The observatory is part of Boston College, a Jesuit university, and signs of Catholic influence abound. To one side of the observatory building, for example, stands a statue of the Virgin Mary. Yet among all the world's statues of the Virgin Mary, this one has a unique feature: at its base lies a slab of rock in which are imprinted the fossilized footsteps of dinosaurs.

Jesuits have a long history of engagement in science, explains John Ebel, director of the observatory. They were the first Catholics, for example, to embrace Darwin's theory of evolution. The Weston Observatory, devoted to seismological observation and research, is very much in the Jesuit tradition. In fact, Ebel is the first non-Jesuit to head the institution; he took over a few years ago from Father James Skehan, S.J., an expert on the geological history of the Atlantic seaboard.

Although the East Coast is not the most seismically active region in North America, it does have the longest history of European presence and thus the longest written record of North American earthquakes. Jesuits and other clerics were the authors of many of these reports, which Ebel has collected and analyzed.

One of the strongest shocks to strike eastern North America in the historical period was an earthquake that rocked Quebec province on February 5, 1663, causing major landslides and liquefaction along the St. Lawrence River. "It began half an hour after the close of Benediction," wrote a Jesuit missionary, "and lasted about the length of two *misereres*." Ebel is Catholic, but not Catholic enough, evidently, to know how long it takes to recite a *miserere*. By asking one of his Jesuit colleagues, however, he determined that it takes about 90 seconds, so the earthquake lasted about 3 minutes. It caused panic as far away as Boston. Ebel estimates that the earthquake had a magnitude of about 7.0.

Boston was much more severely affected by an earthquake nearly a century later, on November 18, 1755, just 17 days after the great earthquake that destroyed Lisbon, Portugal. One letter, written by a resident of Boston a few days later, included the following statement:

> Your Lordship will excuse me for just mentioning a most awakening call of Providence to us especially in this Lane—On the 18th Instant about 25 minutes past four in the morning a very severe shock of an earthquake was felt in this Town and I suppose all over New England. A great part of the Houses in the Town were damag'd, many chimneys thrown down, and roofs beat in with the fall of the Bricks; but thro' the mercy of God, no Man's Life lost: It is esteem'd the severest shock by far that was ever felt in New England.

As with so many other earthquakes, the greatest damage was in areas of landfill. "In some places," according to the *Boston Gazette,* "especially on the low, loose ground, made by Encroachments on the

Harbour, the Streets are almost covered with the Bricks that have fall-en." According to the *Boston Weekly News-Letter,* "Such a violent con-cussion was never known in this Part of the World since inhabited by the English, as plainly appears by the Devastation it has occasion'd, which, to this Town, seem'd to be greatest in the Low Land near the Docks and Wharffs."

The region most strongly affected by the 1755 earthquake was the coastal strip from Boston north to Portsmouth, New Hampshire, although some damage was reported as far north as Portland, Maine, and as far west as Northampton, in western Massachusetts. The earth-quake was also felt far out to sea, as one entry in the *Boston Gazette* attested:

> By a person which came in, Capt. Burnam, who arrived at Marblehead from Cadiz last week, we learn that they felt the above shock 70 leagues [242 statute miles] E of Cape Ann, at half past four, but con-cluded that they ran foul of a Wreck, or got upon a Bar, but on throw-ing over the Lead, found they could not sound in 50 fathom of Water, and continued ignorant of what it was till Morning, when to their great Surprise, they saw a vast number of Fish, large as well as small, floating on the Water dead, when they concluded that it could be noth-ing but an Earthquake, and were informed it was so, as they were going into Harbour.

The 1755 earthquake is referred to as the Cape Ann earthquake, but it is generally believed that the earthquake was centered some distance, perhaps 40 miles, out to sea. The killing of fish by earthquakes has been reported on other occasions; it is probably a result of the direct con-cussive effect of the strong compressive primary waves.

The greatest historical earthquake in the eastern United States struck Charleston, South Carolina, in 1886. On August 27 and 28, a series of precursory "explosions" rattled the village of Summerville, 22 miles northwest of Charleston. The mainshock came at 9:50 P.M. on August 31. One hundred and two buildings in Charleston were com-pletely destroyed, and nearly 14,000 chimneys fell. About 60 people were killed.

Unlike Boston in 1755, where most buildings were wood-framed with brick chimneys, Charleston in 1886 boasted a large number of all-brick buildings. This was in part a consequence of a great fire that destroyed much of the city in 1838; after this event, a law was passed forbidding the construction of wooden buildings. Furthermore, the

Buildings in Charleston, South Carolina, damaged by the 1886 earthquake. (From contemporary report by C. Dutton, USGS)

Locomotive derailed by the Charleston earthquake. (From contemporary report by C. Dutton, USGS)

new brick buildings were constructed with inferior mortar that failed to hold the bricks together during the earthquake. Old prerevolutionary brick buildings, built with mortar made from local oyster shells, survived the shaking with only minor damage.

The perils of trying to leave a brick building during an earthquake are illustrated by the following account, written by a man who was staying at the Charleston Hotel:

> At the very first shock the lights in the house all went out, and I was in perfect darkness. Then the plastering began to fall. It flashed through my mind that I should endeavor to get out of the house, and I got out into the corridor and groped my way, in utter darkness, amid falling plaster. Other people in the house were likewise groping their way out. When I reached the ground floor I could see a little from the reflection from the street lights. The air was filled with plaster dust. All around was a terrible roaring and moaning sound, and the din was heightened by the falling of timbers. I found the front door of the house closed, a fortunate thing for me, as it saved my life. It took me a moment to find the knob, and as I was looking for it tons of brick fell down from the upper part of the house in front of the door. If I had found the door open I would have been buried under the bricks. I ran out through the heaps of fallen bricks and fell twice in getting to the middle of the street. There I remained terror-stricken.

Several witnesses who were out of doors at the time of the earthquake gave reports of the enigmatic visible waves, similar to Audubon's account in the previous chapter. Here is one account:

> The street was well lighted, having three gas lamps within a distance of 200 feet, and I could see the Earth waves as they passed as distinctly as I have a thousand times seen the waves roll along Sullivan's Island beach. The first wave came from the southwest, and as I attempted to make my way towards my house, about 100 yards off, I was borne irresistibly across from the south side to the north side of the street. The waves seemed then to come from both the southwest and northwest and crossed the street diagonally, intersecting each other, and lifting me up and letting me down as if I were standing on a chop sea. I could see perfectly and made careful observations, and I estimate that the waves were at least two feet high.

Although Charleston sustained much damage, the most severe shaking was experienced north of the city in the region around Summerville, where the foreshocks had been felt. Here, even wood-

Damaging seismic waves spread farther in the eastern United States than in the West. As evidence, note the dramatic difference in the size of the regions subjected to severe shaking in four famous historical earthquakes. (Adapted from a USGS circular)

frame houses were wrecked, and fissures and sand-blows tore up the ground. The tracks of the South Carolina Railroad buckled in many places, and at least one locomotive was overturned. Because the telegraph wires were also broken in this area, Charleston was cut off from the world for a day or two. Only later did the residents find out that their earthquake was felt from Canada to New Orleans and from Arkansas to Cuba.

The great extent of the territory that was shaken during this earthquake confirms once again how efficiently seismic waves travel in the rock underlying the eastern and central United States, as compared with California. We can put figures on this difference by estimating the extent of moderate shaking (sufficient to cause slight to moderate damage in well-built ordinary structures). During the 1906 San Francisco earthquake, which had an estimated magnitude of 7.7, moderate shaking affected an area of only about 20,000 square miles. (Even directly across San Francisco Bay, in Berkeley and Oakland, damage was minor.) The Charleston earthquake released far less energy than the San

Francisco earthquake—its estimated magnitude was 7.0—but moderate shaking affected about 50,000 square miles. The most powerful of the New Madrid shocks—an estimated 8.1—caused moderate shaking over more than 200,000 square miles!

The Origin of East Coast Earthquakes

If seismologists had only East Coast earthquakes to go on, it's unlikely that they would yet have figured out the connection between earthquakes and the rupture of geological faults. For the Atlantic states have plenty of earthquakes, and they have plenty of faults, but putting the two together has proved vexingly difficult.

According to Ebel, there are a number of reasons for this. First, earthquakes in the eastern United States are rarely accompanied by rupture of a fault to the ground surface. In fact, there has been no historical earthquake in this region for which surface rupture of a fault has been verified.

Second, there are almost too many faults to choose from. The crust along the Atlantic seaboard is riddled with faults, most of them inactive, that were generated during this region's long and tortured geological history. Because the shaking from even moderate earthquakes in the region extends over much wider areas than it does during comparable earthquakes in the western United States, there is not usually a single obvious candidate fault for the origin of a particular earthquake.

Another problem, which is specific to the northern Atlantic states, has to do with the history of glaciation. Until about 10,000 years ago, glaciers extended as far south as Cape Cod and Long Island. The moving ice either ground down or deposited debris on the underlying land surface, thereby erasing all likely signs of previous earthquakes—scarps, small offset features, and so forth. Thus, the visible record of earthquakes is a brief one—in a part of the world where faults probably rupture at very long intervals.

If, however, we assume that earthquakes on the East Coast, as elsewhere, are indeed the consequence of fault ruptures, then we are up against the same question that we faced in the case of the New Madrid seismic zone: Why are faults slipping? Most of the East Coast is thousands of miles from the nearest plate boundary, and it should therefore be relatively immune from the dislocations that characterize the boundary zones.

Before plate tectonics came of age, glaciation was considered the leading candidate for the origins of earthquakes in the Northeast. During the glacial epoch, the great ice sheets weighed down the underlying continental crust. Beyond the limits of the ice, the crust bowed upward in compensation. Then, when the ice melted, these deformations of the crust began to reverse themselves: eastern Canada and the New England states began to rise, while the crust farther to the east and south began to sink. These movements are continuing today, and it seemed reasonable to believe that they are responsible for current seismicity.

According to Ebel, however, the current distribution of stress in the crust does not fit well with an origin in glaciation-induced warping. If glaciation were the cause, the principal axis of crustal stress in the region would be from northwest to southeast—the axis of most rapid change from upward to downward deformation of the crust. But on the basis of the analysis of seismic waves from many small earthquakes in the region, as well as from measurements of stress obtained in boreholes, it appears that the major stress axis in the region is approximately east-west. This is the axis of motion of the North American Plate relative to the Mid-Atlantic Ridge. Thus, it seems probable that the major force contributing to the stress field is simply the push being exerted at the Mid-Atlantic Ridge. This push may be resisted by drag on the underlying mantle or by the opposing force exerted by the Juan de Fuca and other plates at the subduction zone in the Pacific Northwest.

Although the regional patterns of stresses related to the plate motions are probably the ultimate cause of seismicity, there are doubtless local features that influence the distribution of earthquakes. In Chapter 10, for example, we mentioned the existence of a failed rift in the New Madrid area that may be part of the reason for the occurrence of great earthquakes there. There are probably local features that influence the occurrence of earthquakes on the East Coast, too. The geological history of the area, which includes two episodes of continental collision and rifting over the past 600 million years, has left abundant scars on the subterranean landscape: rift zones, plutons, bits of Africa that somehow ended up in Massachusetts, ancient faults that have been folded into squiggles. The difficulty lies in pinpointing the particular kind of defect that attracts earthquakes.

Pradeep Talwani, a geophysicist at the University of South Carolina at Columbia, believes that large intraplate earthquakes are particularly likely to occur at locations where fault systems (or other preexisting zones of weakness) intersect each other. Such intersections, he thinks, are places where stresses are concentrated to much higher levels than elsewhere. Talwani believes that the intersection of two fault systems (the Woodstock and Ashley River fault zones) in the region northwest of Charleston is a factor predisposing the region to large earthquakes.

Because the overall level of seismicity is so low, neither of the these two zones is delineated as well as the fault zones at New Madrid. Nor are they well delineated underground by seismic reflection studies or other techniques. But the Ashley River fault zone does seem to have influenced the topography of the ground surface: several of the rivers that cross it are deflected northward—probably by upward warping of the ground on the western (hanging-wall) side of the fault zone—in the same manner as the Mississippi has been deflected northward by cumulative uplift on the Reelfoot fault.

Although there have been no trenching studies in the Charleston area, comparable to the work on the Reelfoot fault, much evidence about previous earthquakes has come from the study of sand-blows and other liquefaction features. This work has been done primarily by Stephen Obermeier (of the USGS in Reston, Virginia), by David Amick and Robert Gelinas (corporate geologists in Greensboro, North Carolina), and by Talwani. It turns out that earthquakes strong enough to cause widespread liquefaction have occurred at least five times near Charleston over the past 4000 years. The two most recent events prior to the 1886 earthquake took place approximately 600 and 1300 years ago. These data suggest that large earthquakes may happen in the Charleston area about every 500–600 years. There is also evidence for one large earthquake that seems to have been localized to the northern part of South Carolina; it took place about 1700 years ago.

To get some idea of whether South Carolina has had more than its share of earthquakes over the centuries, Amick and Gelinas traveled up and down the Atlantic seaboard, from southern New Jersey to the Georgia-Florida state line, looking for evidence of soil liquefaction. They identified more than 1000 locations outside the state of South Carolina where soil conditions were appropriate for earthquake-induced liquefaction, but aside from one site just north of the border between North Carolina and South Carolina, they found no sand-blows or any other evidence that liquefaction had ever taken place. Thus, South Carolina's history of large earthquakes does seem to be

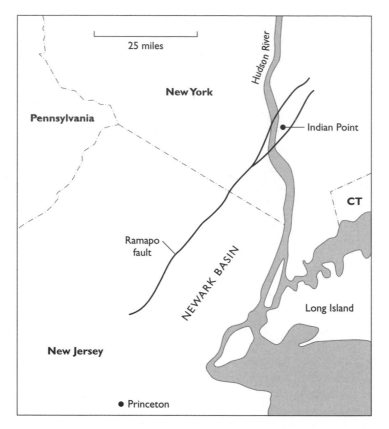

The Ramapo fault is an ancient fault that may pose a seismic threat to New York City. (Adapted from an illustration by A. Kafka and others)

unique, and most of this history seems to involve earthquakes near Charleston. As a result of these and other studies, a bull's-eye of increased seismic hazard has been drawn around Charleston in the national seismic-hazard maps (see Color Plate 18), similar to the one drawn around New Madrid.

The city of New York also earns a seismic bull's-eye, although not nearly as intense as the one drawn around Charleston. Damaging earthquakes struck the city in 1737 and in 1884. Although neither of these events was in the same league as the Cape Ann or Charleston earthquakes, they were still significant. In the 1884 earthquake, the most severe shaking was at the western end of Long Island, but minor damage was reported from as far afield as eastern Pennsylvania and central

Connecticut. Noticeable but nondamaging earthquakes (up to a magnitude of about 3.0) strike the New York City area on a fairly regular basis. On one August day in 1938, three earthquakes occurred in the space of four hours. They had magnitudes of 3.9, 4.0, and 3.7 and were centered about 30 miles south of the city, in New Jersey.

Seismologists at the Lamont-Doherty Earth Observatory in Palisades, New York, have focused their interest on a particular fault—the Ramapo fault—as a potential source of some of this seismic activity. The Ramapo fault lies about 20 miles northwest of the city; it runs from the area west of Newark, New Jersey, to the Hudson River south of West Point. The fault is extremely ancient: it was probably active in Precambrian times, more than 600 million years ago, and has been reactivated several times since then. About 100 million years ago, it functioned as a normal fault that bordered the Newark Basin, a depression produced during the process of continental rifting that lead to the opening of the Atlantic Ocean. If the Ramapo fault is currently active, it would be expected to act as a reverse fault, but excavations have not revealed evidence of reverse motion on the fault in the past many millions of years.

Interest in this fault became more intense in the 1970s, when the Indian Point nuclear power station was constructed within half a mile of the fault, where it crosses the Hudson. The Lamont-Doherty scientists determined that a number of small earthquakes had originated in the vicinity of the Ramapo fault and tentatively attributed them to movement on the fault. But further studies cast doubt on this conclusion. Nano Seeber of Lamont-Doherty mapped the distribution of the tiny aftershocks of some of these earthquakes and concluded that the earthquakes were caused by slippage on relatively minor faults oriented at right angles or obliquely to the Ramapo fault. These cross-faults are probably too small to be capable of causing damaging earthquakes. Whether the Ramapo fault itself is a serious candidate for producing a damaging earthquake remains an unanswered question.

The Ramapo fault is just one part of an extended, ancient fault system that rims Mesozoic (250 to 65 million-year-old) basins in the mid-Atlantic states. At the southwestern end of the Newark Basin, the "border fault" runs near Reading, Pennsylvania. On January 15, 1994, an earthquake near (but not on) this fault caused $2 million in damage to a small area west of Reading. Although the earthquake's magnitude was only 4.6, the rupture occurred at very shallow depths (between the surface and a depth of about 1.2 miles). For this reason the shaking was localized to a small area, but within that area it was disproportionately severe. This Cacoosin Valley earthquake was one of the most damaging earthquakes to strike the East Coast in recent times, but in terms of media attention it was overshadowed by the Northridge earthquake that struck Los Angeles just two days later.

Vulnerability and Preparedness

Because of a general perception that seismic hazards on the East Coast are low, there has been correspondingly little attention paid to mitigation. Even in South Carolina, seismic preparedness is minimal, according to Talwani. The situation in other states is no better. The city of Boston is perhaps more vulnerable to earthquakes today than it was in 1755. During the nineteenth century, developers sliced off the top of Beacon Hill and hauled it in carts to the Back Bay. The south side of the bay was filled in, and unreinforced masonry row houses were constructed on the new land. As if this combination—unconsolidated land-fill and unreinforced masonry—were not sufficiently hazardous, John Ebel mentions yet a third problem in the Back Bay district. The buildings were constructed on wooden pilings driven into the fill. These pilings might have remained in good condition indefinitely, but a lowering of the water table has exposed their upper ends to the air, and this in turn has allowed many of them to rot. Thus, some of the buildings in the Back Bay are not securely attached even to the untrustworthy ground they were built on. A repeat of the Cape Ann earthquake would put the Back Bay at serious risk.

According to Ebel, the scientific community also pays too little attention to earthquake hazards in the Northeast. The Weston Observatory is the nerve center of the Northeastern United States Seismic Network, which collects and analyzes data from more than 100 seismic stations in the region. Federal funding for the network (which had come from the Nuclear Regulatory Commission) was cut off in 1992, forcing Ebel to reduce his staff and to run a bare-bones service.

Ebel, who obtained his Ph.D. at Caltech, is well aware of the resources that are devoted to seismological research in California, and he feels that the eastern states don't get a fair share of the pie. He also suggested that too many resources are invested in testing what may be unrealistic theories. As an example, he cited the expensive, decades-long effort to monitor the San Andreas fault at Parkfield. "The Parkfield earthquake prediction experiment is a failure," he told us emphatically. "The earthquake didn't occur." When we suggested that the Parkfield earthquake, though tardy, might happen before this book comes out, Ebel smiled and predicted that God would now defer the date of the earthquake even further. "Being a Catholic, and being at a Jesuit university, I have the firm opinion that God has a sense of humor, and I think this is one of the ways He shows it."

Having reached the East Coast, we seem to have come to the end of our cross-country trip. But just as cartographers tuck the Hawaiian Islands into some incongruous but convenient location when they construct maps of the United States, so we have taken the same liberty and deal with Hawaii out of sequence. We have both scientific and literary reasons for doing so. All the regions we have discussed so far have a unifying feature: they are composed of continental crust. The Hawaiian Islands, in contrast, are oceanic. Also, to close our book with the East Coast, where seismologists fall asleep for lack of excitement, would be to invite our readers to do the same. Hawaii is the very opposite—there, tectonic processes take place at such a rate as to make the very islands seem alive.

12

PELE'S WRATH: THE VOLCANOES OF HAWAII

In the year 1790, Kamehameha, chieftain of the northernmost district of the Big Island of Hawaii, invaded and subdued the neighboring island of Maui. Flushed with success, he developed the ambition to conquer all the Hawaiian Islands, and he advanced to the next large island, Molokai. While planning his next step, the invasion of Oahu, he received word that Keoua, chieftain of the southern Kau district of the Big Island, had taken advantage of his rival's absence to ravage the island's northeast coast, including a site sacred to Kamehameha's war god Kukailimoku. Enraged, Kamehameha headed back to the Big Island, and Keoua's army beat a hasty retreat towards Kau. Their route took them by the summit caldera of Kilauea, the 4000-foot-high volcano

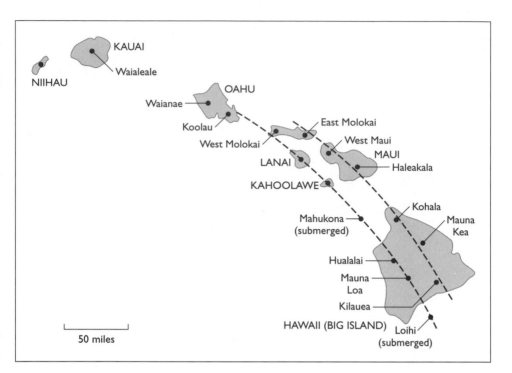

The major volcanoes of the Hawaiian Islands. Note the tendency for the volcanoes to line up in two parallel rows (dashed lines).

near Hawaii's southeast coast. The area had been struck by a series of intense earthquakes over the previous few days. Perhaps seeking divine protection, the army encamped near the temple of Oalaluao, sacred to Pele, the goddess of fire. The events that followed were later recounted to Western missionaries:

> During the night the eruption began by throwing out cinders and even heavy stones, the whole accompanied by the glare of molten lava, thunder and lightning. Fear struck, the party in the morning did not dare go on, but spent the night in making offerings to Pele, but as on the next two nights there were similar disturbances they at last set out in three divisions.
>
> The company in advance had not proceeded far, before the ground began to shake and rock beneath their feet and it became quite impossible to stand. Soon a dense cloud of darkness was seen to rise out of the crater, and almost at the same instant the electrical effect upon the air was so great that the thunder began to roar in the heavens and the lightning to flash. It continued to ascend and spread abroad until the whole region was enveloped, and the light of day was entirely excluded.

The darkness was the more terrific, being made visible by an awful glare from streams of red and blue light variously combined that issued from the pit below, and being lit up at intervals by the intense flashes of lightning from above. Soon followed an immense volume of sand and cinders which were thrown in the high heavens and came down in a destructive shower for many miles around. Some few persons of the forward company were burned to death by the sand and cinders, and others were seriously injured. All experienced a suffocating sensation upon the lungs and hastened on with all possible speed.

The rear body, which was nearest the volcano at the time of the eruption, seemed to suffer the least injury, and after the earthquake and shower of sand had passed over, hastened forward to escape the dangers which threatened them, and rejoicing in mutual congratulations that they had been preserved in the midst of such imminent peril. But what was their surprise and consternation, when on coming up with their comrades of the centre party, they discovered them all to have become corpses. Some were lying down, and others were sitting upright clasping with dying grasp their wives and children, and joining noses (their form of affection) as in the act of taking a final leave. So much like life they looked, that they at first supposed them merely at rest, and it was not until they had come up to them and handled them that they could detect their mistake. The whole party, including women and children, not one of them survived to relate the catastrophe that had befallen their comrades. The only living thing they found was a solitary hog. . . . In those perilous circumstances, the surviving party did not even stay to bewail their fate, but leaving their deceased companions as they found them, hurried on and overtook the company in advance at the place of their encampment.

Kamehameha went on to become Kamehameha the Great, first hereditary king of all the Hawaiian Islands. Keoua survived Pele's anger, only to become the first human sacrifice at Kamehameha's new temple to Kukailimoku, the war god. And the footprints of Keoua's ill-fated soldiers and their families, left in a thin layer of whitish volcanic deposits, still remain on Kilauea's southwestern slopes, alternately hidden and exposed to view as the wind blows 200-year-old black ash across them.

The USGS's Hawaiian Volcano Observatory, perched on the high northern rim of Kilauea's summit caldera, has inherited some of the functions of the ancient temple of Oalalauo. Where once the soothsayer Kamakaakaakua interpreted Pele's capricious moods, Dr. Christina Heliker and her colleagues pore over the digital output of seismographs and tiltmeters, map the advancing lava flows, and sample gases issuing from the latest vents. Having completed their divinations, they issue their oracles to an anxious populace.

Anxious, that is, that there be some kind of eruption, for Kilauea, active off and on throughout historical times, has become the islands' third most-visited attraction. It has developed a reputation as a spectacular yet tourist-friendly giant: a volcanic petting zoo whose fire fountains and lava flows, should they turn surly, can be escaped simply by walking away. So confirmed in this view are some Hawaiians that they build their homes right in the most likely paths of future eruptions—in what Heliker calls Hazard Zone One areas. If the lava comes their way, they seem to think, they can halt it with garden hoses. Sometimes that even works. Or if worse comes to worst, they will have time to load their houses onto flatbed trucks and move to the next valley. Sometimes that works, too.

But Heliker knows that Hawaiian volcanoes, especially Kilauea and her giant neighbor Mauna Loa, are capable of unleashing death and violent destruction. If you take away all the peaceful eruptions, she told us during a recent visit, you're still left with an impressive roster of explosive ones. And even the peaceful eruptions have the potential to take lives and cause enormous economic disruption. Heliker hopes that her long and intimate acquaintance with the fire goddess—she has been a member of Kilauea's monitoring team continuously since 1984—will help her to provide timely and accurate forecasts of future eruptions and to predict the direction and speed of lava flows once an eruption begins. Even more important, Heliker's mission is to educate—to help Hawaiians understand the volcanoes and thus make wiser decisions about land use than they have in the past.

Beyond this practical mission, the scientists at Kilauea see their observatory as akin to the astronomical observatory on Mauna Kea, 30 miles to the north: a set of telescopes aimed not toward the heavens, but toward the even more mysterious interior of our own planet. Kilauea, and the Hawaiian volcanoes in general, offer unparalleled opportunities to study the Earth's never-ending evolution. In Hawaii, rock from a depth of 1000 miles or more comes to the surface in copious volumes. On a good day, one can sit and watch mountains rising. What's more, the entire chain of Hawaiian islands, and the submarine mountains to the northwest of them, form an album of snapshots that vividly illustrate the life history of an oceanic volcano, from its stormy youth through its majestic prime to its crabbed old age.

The Track of the Hawaiian Hot Spot

According to Hawaiian legend, Pele first established her home on the northernmost island of Kauai and then moved southeastward, island by island, mountain by mountain, until she settled at last on Kilauea's fiery summit. It seems that the native Hawaiians knew, in some sense, that the Hawaiian volcanoes are an age-based series—that Kilauea is young and Kauai old. Not that they were around to witness this evolution, of course: the few hundred years that humans have inhabited the islands is a mere instant in geological time, even in the frenetic pace of Hawaii's geological evolution. They must have understood that the landmasses of the northern islands are the eroded, sunken, and disintegrating remains of giant shield volcanoes, like those that now dominate the archipelago's southernmost reaches.

Modern science has confirmed and fleshed out this understanding. In particular, scientists can do two things that the Hawaiians, for all their insight, could not do. First, at the flick of a digital switch, they can empty the whole Pacific basin of its water, exposing the surface of the oceanic crust to view—not just a sprinkling of palm-fringed islands but the endless sediment-smooth expanses of the Pacific Plate and the arrays of submarine mountains, or *seamounts,* that punctuate it. And second, as described in Chapter 1, scientists can date rocks by the potassium-argon technique or other methods. Thus, they can supply numbers where the Hawaiians had only intuition.

In the absence of the obscuring seawater, the Hawaiian-Emperor volcanic chain stands out as one of the most striking linear features on the Earth's surface. It extends about 3600 miles across the Pacific Plate, starting at the Kurile Trench (a subduction zone off Siberia's Kamchatka Peninsula) and ending just south of the Big Island of Hawaii. The chain has two bends. One is near the beginning of the chain; the other, much more striking, is nearly halfway along the chain. This second bend is called the Hawaiian-Emperor bend: the part of the chain to the northwest of it is called the Emperor seamount chain, and the part to the southeast is called the Hawaiian Island chain.

The mountains of the chain are progressively taller from northwest to southeast. The mountains of the Emperor chain are entirely submarine. Of the mountains of the Hawaiian chain, some are submarine; others graze the ocean surface, forming the tiny, scattered Leeward Islands; and still others—those at the southeast end of the chain—rise

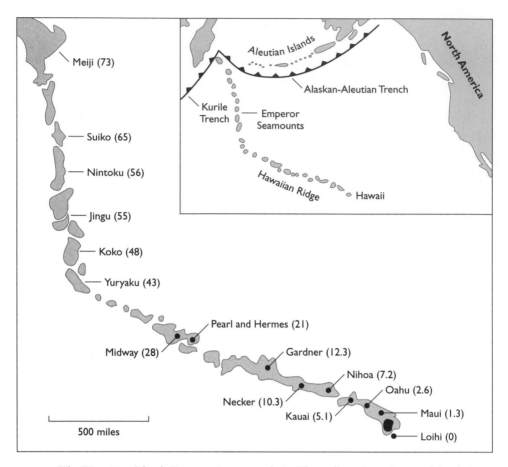

The Hawaiian Island–Emperor Seamount chain. The atolls and seamounts of the chain are shown by their 2-kilometer (1.2-mile) depth contours. The numerals are the potassium-argon ages of rock samples, in millions of years. (Adapted from an illustration by D. Clague and G. B. Dalrymple, USGS)

well above the water, forming the major Hawaiian Islands. The two tallest mountains of the entire chain are Mauna Kea (13,796 feet above sea level) and Mauna Loa (13,677 feet), both on the Big Island. Measured from their foundations on the Pacific Plate, these two mountains are about 33,000 feet tall—the world's highest. Mauna Loa is also the world's largest volcano; its volume of about 18,000 cubic miles is greater than that of all the mountains of California's Sierra Nevada combined. (Still, Mauna Loa is less than half the height and one-hundredth the volume of the great volcano on Mars, Olympus Mons.)

We already mentioned, while discussing Yellowstone in Chapter 9, how J. Tuzo Wilson and, later, W. Jason Morgan developed the notion of a hot spot as the mechanism for the formation of the Hawaiian-Emperor chain. A stationary mantle plume, they proposed, currently underlies the Big Island and is responsible for volcanism there. The motion of the Pacific Plate over the mantle plume has carried older volcanoes, produced by the same hot spot, in a northwestward direction. Thus, the Hawaiian-Emperor chain marks the track left by the Hawaiian hot spot, just as the extinct volcanoes of the Snake River Plain mark the track left by the Yellowstone hot spot. Because the oceanic lithosphere is so thin and so uniform, the Hawaiian track is far more obvious and easy to interpret than the Yellowstone track, which underlies the thicker and more variegated continental lithosphere.

The major evidence for the hot-spot theory of the origin of the Hawaiian-Emperor chain comes from the dating of rock samples at various sites along the chain. Samples of rocks from Suiko seamount in the Emperor chain are 60–65 million years old. Samples from Yuryaku seamount, at the big bend between the Emperor and Hawaiian chains, are 43 million years old. Among the main Hawaiian Islands, ages vary from about 5.1 million years old (for samples from the northern island of Kauai) to newborn (for samples from Kilauea on the Big Island and from Loihi seamount).

From the dates of the rock samples along the chain, it is a straightforward matter to calculate the past direction and speed of motion of the Pacific Plate with respect to the hot-spot reference frame. Before about 73 million years ago, the plate was moving approximately northwestward. This early period is not well represented in the Emperor chain, because the oldest seamounts (and presumably the flood-basalt province that was the starting point of the track) have been carried down into the Kurile subduction zone. But it is well documented by other island chains produced by other hot-spot tracks in the Pacific Ocean.

About 73 million years ago, the direction of motion of the Pacific Plate turned by about 30 degrees clockwise, producing the first bend in the hot-spot track. Subsequently, it moved at about 70 millimeters (2.8 inches) per year in an approximately north-northwestward direction, laying down the Emperor chain. About 43 million years ago, the direction of plate motion veered through about 60 degrees counterclockwise, producing the second, major bend in the chain. Since then it has moved at about 86 millimeters (3.4 inches) per year in an approximately west-northwestward direction, laying down the Hawaiian chain.

A number of scientists have wondered what caused the 60-degree shift in direction 43 million years ago. Of course, the Earth's tectonic plates are engaged in such an intimate dance that events at one location—the appearance or disappearance of a mid-ocean ridge or a subduction zone, for example—may ultimately affect plate motions on the other side of the planet. Among other events that happened in that general time period were the collision of India and Asia, the separation of Australia and Antarctica, and the development of new subduction zones along the southwestern edge of the Pacific Plate. Exactly how these events might have led to a change in the plate's direction of motion must remain a matter of speculation, however, until a detailed dynamic model of all the Earth's plate motions has been constructed.

One interesting feature of the volcanoes in the Hawaiian-Emperor chain is that they seem to be arranged not in a single line, but in two or more parallel rows. At the southern end of the Big Island, there are two currently active volcanoes: Mauna Loa to the west and Kilauea to the east. These are the youngest members of two rows. Behind Mauna Loa comes Hualalai and several other volcanoes reaching back to Koolau on the island of Oahu. Behind Kilauea comes Mauna Kea and other volcanoes reaching back to the eastern volcano on Molokai. The two rows are about 25 miles apart. The older seamount volcanoes to the northwest also seem to be arranged in two or three rows. The reason for this arrangement is the subject of lively debate. According to one theory, there is actually more than one hot spot. In another theory, put forward by Phil Ihinger of Yale University, the rowlike arrangement is caused by sideways shearing motions of the plume as it rises through a countercurrent in the mantle beneath the lithosphere.

Volcanic Infancy: Loihi

Because the Hawaiian-Emperor chain presents an evolutionary time series, we can visualize the life history of a typical Hawaiian volcano by taking a look at some of the individual volcanoes, starting at the southeast and moving progressively to the northwest. The starting point of this series, however, is not the Big Island of Hawaii, but a seamount 15 miles off the Big Island's southeast coast.

Loihi, as this seamount has been named, was originally discovered in the 1950s during acoustic surveys of the ocean bottom. It is an oval-shaped mountain (*loihi* means "long") that is oriented north-south. The northern end of the seamount sits on top of the southern, submarines slopes of Mauna Loa, while the southern end sits directly on the ocean floor. Its height is 10,000 feet, and its summit is 3179 feet below the ocean surface.

Originally, Loihi was thought to be just another of the decaying volcanic edifices that litter the Pacific basin, most of which have about the same age as the plate on which they stand. But several later observations served to cast doubt on this assumption. Earthquakes, apparently originating from Loihi, were recorded from time to time by seismographs on the Big Island. Samples of rock dredged up from the summit region turned out to be relatively young pillow lavas—the kind of lava flow typically formed when basalt erupts underwater. Photographs taken with cameras lowered from boats showed extensive, fresh-looking lava flows, as well as what seemed to be hydrothermal deposits—layers of minerals deposited on the seamount's surface from hot brine rising through fissures and cracks in the volcano. Thus, it gradually became apparent that Loihi is actually the youngest Hawaiian volcano, so youthful that it has not yet reached the ocean surface. From its position, Loihi appears to be the next volcano in the row to which Mauna Loa belongs.

In the mid-1980s, scientists at the University of Hawaii began making visits to Loihi, with the help of a submersible, *PISCES-V.* One of these explorers, Alexander Malahoff, described to us what they found. Most of the slopes of the seamount were covered by *talus*—irregular blocks of rock that reached their present position by rolling or sliding downhill. Parts of the sides of Loihi were indented, as if large sectors of the seamount had fallen into the abyss. Thus, landsliding seems to be an important process in shaping the volcano. This makes sense, because lava erupted underwater cools rapidly and thus forms steep, unstable slopes, quite different from the gently sloping flows produced by the same kind of lava when erupted on land.

In the summit area, Malahoff and his colleagues found all the signs of an active volcano, short of an actual eruption. There were two large pit craters, about 300 feet deep, as well as several volcanic vents that ranged above the general topography. One of the most prominent of these, which was given the name Pele's Vent, was the scene of extensive hydrothermal activity: the rocks were covered with metal sulfides, such as pyrite (fool's gold), that had precipitated out of solution when the hot volcanic fluids were cooled by seawater, and with colonies of bacteria, not unlike those in Yellowstone's hot pools, that used the dissolved chemicals as an energy source. Water samples taken in these vent areas contained high concentrations of methane, helium, and carbon dioxide, indicating that volcanic gasses were being released into the seawater.

Leading away from the summit to the north and south were two rift zones—regions of recent volcanic activity that appeared to stretch down toward the base of the volcano. The mature volcanoes of the Hawaiian Islands also have such rift zones, which play a major role in the volcanoes' eruptive life. Evidently, the summit crater–rift zone system develops early in the life of a Hawaiian volcano.

Intrigued by these signs of current volcanism, the University of Hawaii scientists conceived a plan to "wire" Loihi: to lay a submarine fiber-optic cable from the Big Island to Loihi, which then could be connected to seismographs, pressure meters (which operate like tilt-meters on land, registering deformation of the volcanic edifice), and other devices that were to be installed on Loihi's summit. The idea was to have a permanent, remotely monitored volcanic observatory on Loihi. Unfortunately, this project became ensnarled in planning delays, and there were no functioning instruments on Loihi on July 16, 1996, when a major volcanic event began.

The first that scientists knew about this event was a series of earthquakes, picked up by seismographs located on the Big Island. By the following day, there had been 72 shocks. After a day of calm, the earthquakes began again. By August 9, more than 4000 earthquakes had been detected by the land-based recording stations, making it the biggest earthquake swarm ever recorded in Hawaii—the land of earthquake swarms.

The scientists rushed to their boats, and on August 7 Fred Duennebier, chief scientist of the Hawaii Underwater Geo-Observatory, descended to the summit of Loihi in *PISCES-V*, along with two colleagues. As they descended, sailors on the mother ship waiting overhead could feel the seismic shocks that were being transmitted through 3000 feet of seawater and into the hull of their ship. Shipboard scientists also picked up strange noises—bangs, pops, and grinding sounds—with the aid of microphones hung from free-floating buoys.

Duennebier's party was both fascinated and alarmed by what it saw. The water was thick with stirred-up sediment, bacterial-mat debris, and other material that reflected *PISCES-V*'s searchlights back in a blinding glare. Sometimes they could see no more than 5 feet in front of the submersible, and they had to navigate with extreme caution. Furthermore, they soon realized that the landscape below them had changed drastically. The familiar landmark of Pele's Vent had completely disappeared; it was replaced by a giant crater filled with sediment-clouded water. Although they could not see into the murky depths, acoustic instruments on board the mother ship probed the crater and registered a depth of about 1000 feet. An estimated 300 million tons of rock had simply disappeared down into the hole, which was henceforth given the name Pele's Pit.

At Kilauea and the other island volcanoes, the collapse of summit craters often accompanies eruptions along the flanking rift zones, probably because the rift-zone eruption drains magma out of a central reservoir. Dunnebier suspected that a similar mechanism was operating at Loihi, and he went looking for the site of a rift-zone eruption. *PISCES-V* is not able to descend far below the level of Loihi's summit, however, and neither on this dive nor on several subsequent dives was the predicted lava flow located.

Despite this failure, there is reasonably good evidence that there was an eruption in July 1996. First, lava samples collected by Dunnebier were tested with a new radiometric dating procedure that involves measuring the decay of the isotope lead-210 to polonium-210. This technique can date extremely young rocks, and preliminary results suggest that the samples were no more than a few weeks old when they were collected.

Another piece of evidence supporting the idea that there was a rift-zone eruption comes from the analysis of seawater samples. These samples were collected in canisters lowered from ships. One sample of water collected more than 2000 feet below the level of Loihi's summit, but 30 miles away from it, was much more acidic than normal seawater and contained more than twice as much helium-3. Most likely, the canister had sampled a drifting plume of seawater that had taken up volcanic gases emitted during a flank eruption.

The 1996 eruption, if that is what it was, did not raise Loihi's summit. But summit eruptions must also occur sometimes, and by slow degrees the volcano must be rising. Malahoff estimates that the volcano will finally breach the ocean surface about 50,000 years from now, forming a new Hawaiian Island. Later, he predicts, the growing island will fuse with the southern edge of the Big Island, converting it from a triangular to a diamond shape and turning what is now prime beach-front property into a lava-strewn saddle between Loihi and Mauna Loa.

If Loihi is the youngest volcano of the western segment, which includes Mauna Loa, what about the eastern segment? Is there a vol-cano rising from the ocean floor to the east of Loihi, in line with Kilauea and Mauna Kea? Not yet, apparently—surveys of this region of the ocean floor have revealed nothing but rubble that has slumped off Kilauea. Yet eventually, the next member of the Hawaiian chain may show itself there.

Volcanic Adolescence: Kilauea

Kilauea is the best known and most frequently visited of all the Hawaiian volcanoes, because it accessible by car, it is the site of the Hawaii Volcano Observatory, and it has had long periods of almost con-tinuous eruptive activity throughout history. Its most recent eruptions began in 1983 and led to a federal disaster declaration in 1990. All was quiet the last time we visited the volcano, in the spring of 1997, but it has since burst into renewed activity.

Kilauea sits on the southeast flank of Mauna Loa, which in turn sits on the southern flank of Mauna Kea (see Color Plate 20). There are probably places where, if one drilled straight down, one would pass through portions of all three volcanoes. The enormous mass of these three edifices weighing on the oceanic crust has caused the crust to sag downward. Directly beneath the Big Island, the oceanic crust is depressed to far below its normal level. This depression extends side-ways beyond the limit of the island: the seabed flanking the Big Island, as well as the other islands to the north, forms the Hawaiian Deep, where the ocean floor descends 1500 feet below its level elsewhere. Beyond the Hawaiian Deep, the seafloor rises in compensation, form-ing the Hawaiian Arch.

Kilauea is about 250,000 years ahead of Loihi in its evolution. It breached the ocean surface about 200,000 years ago; by 100,000 years ago its summit was about 2300 feet above sea level, and it currently stands 4000 feet above sea level. It is now in the full vigor of its shield-

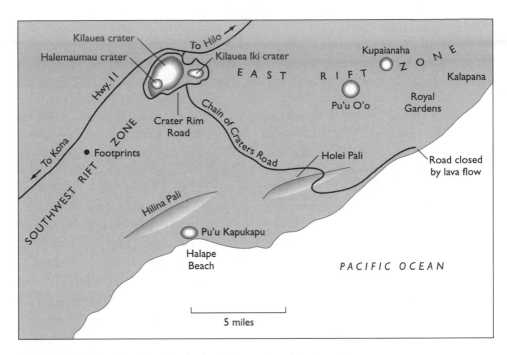

Map of Kilauea volcano showing the localities mentioned in the text.

building phase: so frequent are its eruptions that it takes only about 1000 years for every spot on Kilauea's flanks to be covered with new lava flows—all comprised of tholeiitic basalt. For a while, high points on the volcano's flanks may seem immune to the flows, but the continuing buildup of lava gradually converts high points into low points, and eventually these too are overrun. On Kilauea, all planning is for the short term.

Like Loihi, Kilauea has three main zones of volcanic activity: the summit and two flanking rift zones. One of the rift zones runs down to the southwest, the other to the east. The volcano would probably have a third rift zone, to the northwest, if it were not buttressed on that side by the much higher edifice of Mauna Loa.

All of Kilauea's eruptions appear to be fed from a single magma chamber located a few thousand feet under the summit caldera. Summit eruptions are fed by vertical conduits ascending from the magma chamber to a summit reservoir, whereas the rift-zone eruptions are fed by dikes that branch off from the summit reservoir or the conduits beneath it. For this reason, rift-zone eruptions are sometimes accompanied by summit collapses, as may have happened at Loihi in 1996, or by the drainage of lava pools within the summit caldera.

There seems to be a long-term cyclicity of the eruptions at Kilauea's summit. For a few hundred years, the summit may be a simple broad peak, with lava flowing out of vents near the summit and down the outer surfaces of the volcano. Then, at some point, the summit collapses, forming a caldera. This collapse may be caused by flank eruptions—either outpourings of lava or more explosive events, involving an interaction between rising magma and groundwater that has accumulated over the previous centuries. The 1790 event that destroyed Keoua's army is believed to have been an explosion of this kind, and the Kilauea summit caldera, as we know it today, may have formed around that time. At least two other such explosive eruptions have taken place over the past 3000 years: deposits from the resulting pyroclastic flows and ashfalls blanket at least 100 square miles of the southeast portion of the Big Island.

After these major collapse events, the caldera gradually refills with new lava flows, until finally a new peak has formed. For a few decades after the 1790 event, Kilauea's caldera was about 1000 feet deep—over twice as deep as it is now. It then began refilling, but several episodes of renewed collapse have interrupted the process. In 1868, for example, a magnitude 8 earthquake under Mauna Loa was followed by the collapse of the floor of Kilauea's caldera by as much as 300 feet. From 1905 until 1924, a 500-yard-wide crater within the caldera, named Halemaumau Crater, contained a lava lake whose red-hot churnings mesmerized visitors. Then the lake suddenly drained away, and a series of steam explosions ripped out the walls of the crater, widening it to its present half-mile diameter.

Since that time, activity in Kilauea caldera has been fairly minor. But a smaller crater called Kilauea Iki ("little Kilauea"), just to the east of the main caldera, has seen plenty of action. Kilauea Iki's floor, like that of the main caldera, collapsed following the 1868 earthquake. In November 1959, however, a lava fountain erupted in the southwest wall of the caldera, and it continued erupting until Kilauea Iki was half-filled with molten lava. The lava lake soon crusted over to form what looks, from the caldera rim, like the world's biggest parking lot. But even 20 years after the eruption, temperatures as high as 1150°C were recorded at the bottom of holes drilled into the caldera floor.

Most of the recent activity, however, has been along the east rift zone. After intermittent eruptions during the 1960s and 1970s, the major sequence began in January 1983. A swarm of earthquakes began near the summit and gradually migrated eastward down the rift zone. Volcano Observatory scientists knew the meaning of the earthquake swarm: magma was forcing open a dike in that direction. So they were waiting for the eruption when it finally broke surface, about 8 miles from the summit. As is typical for these events, the eruption initially came from a 4-mile-long fissure—the line along which the dike intersected the ground surface. Low fountains of lava, as well as sheetlike vertical eruptions ("curtains of fire") marked this early stage. But gradually the eruptions became focused at a single vent, about 12 miles east of the summit. At first this vent was named the O vent, because it happened to be located near the letter o in the word *flow* on a map of the area. But when the vent grew into a small mountain in its own right and began sending a fountain of lava 1000 feet into the air, a more respectful name was called for. It was renamed Pu'u O'o ("hill of the o'o"). O'os are extinct birds, 80,000 of which begrudgingly contributed their yellow tail feathers to fashion a cape for Kamehameha the Great. No o'o, alas, ever perched or will perch on Pu'u O'o.

Pu'u O'o erupted about once a month for the next three years and put out about 10 million cubic yards of lava on each occasion. During each eruption, Kilauea's summit caldera sank slightly, only to inflate again before the next eruption. This suggested that magma was steadily rising into the summit reservoir, where it collected until it attained enough pressure to escape through the feeder dike and out of the Pu'u O'o vent.

Then, in 1986, Pu'u O'o missed an eruption. Instead, magma forced open a dike that propagated 3 miles farther down-rift and gave rise to another line of fissure eruptions. Again the eruptions gradually became concentrated in a single vent, which was named Kupaianaha ("mysterious"). This vent simply poured out lava continuously, rather than erupting in periodic fountains. Then, in 1992, the eruptions shifted back to Pu'u O'o, but this time the lava poured out of flank vents rather than jetting skyward. These eruptions, with short intermissions, have continued up to the present time.

The lava flows from the East Rift eruptions, like most flows on Hawaii, are of two kinds—*a'a* and *pahoehoe* (see Color Plates 21 and 22). A'a is all clinkery crags and crevasses—an ugly sight and a torture to walk across, even after it has cooled. Pahoehoe is the opposite—its smooth, loopy oozings please both eye and foot. Keoua's army was following an old pahoehoe flow along the southwest rift zone when it was cut down. It may be that a'a flows to the north and south hampered their escape.

Surprisingly, the same magma gives rise to both a'a and pahoehoe lava flows. Pahoehoe is less viscous than a'a because it contains more gas. It is also often hotter. A pahoehoe flow sometimes turns into a'a flow as it cools and degasses; the reverse is not common. The roughness of the underlying terrain may also play a role: lava flowing across an earlier a'a flow is usually sheared and broken by the rough surface and thus becomes a'a too.

When molten lava pours from a vent, it may initially run downhill in a channel (see Color Plate 21). Sometimes the upper surface of the flowing lava cools and solidifies, roofing over the channel, while the molten lava continues to flow underneath. This is a *lava tube*. Because the roof and sides of the tube act as efficient insulators, the flowing lava may travel for 10 or more miles downhill without an appreciable drop in temperature. Sometimes a piece of the tube's roof collapses, opening up a skylight through which an intrepid visitor can gain a surreal view of the incandescent, subterranean torrent. After the flow ceases, the lava within a tube may drain away, leaving a natural tunnel that can be explored on foot or on hands and knees, depending on its size.

The East Rift eruptions that commenced in 1983 have so far destroyed a total of 181 homes, a church, a store, a visitor center, several historic and sacred Hawaiian sites, and wide expanses of forest and black sand beaches. They have also severed the coast highway. The Royal Gardens subdivision, which lies only 4 miles downslope from the Pu'u O'o vent, has been invaded by lava repeatedly, with the destruction of 76 homes. The most drastically affected area, however, was the subdivision of Kalapana Gardens, a coastal community 20 miles southeast of Kilauea's summit (see Color Plate 23). The assault on Kalapana began in 1986 with the first flows from the Kupahainaha vent. But the coup de grace came in 1990–1991, when new flows fed by a 7-mile-long lava tube wiped out the remainder of the town and continued right on into the ocean. Kaimu Bay, with its famous black sand beach, was completely filled in by the flows, which added 280 acres to the Big Island's land area. Even as we write (early 1998), lava originating at a flank eruption on Pu'u O'o is pouring into the ocean west of the old Kaimu Bay.

Although the surface eruptions are the visible, dramatic aspects of Kilauea's volcanism, about half of all the volcano's magma never makes

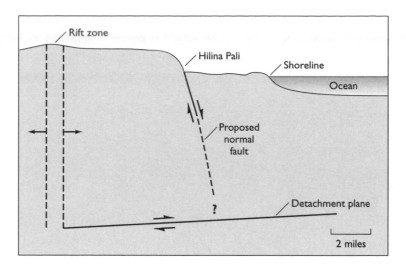

Cross section of the south flank of Kilauea showing a normal fault that may descend from the pali (cliffs) toward the deep detachment plane, which is believed to be the boundary between the Hawaiian volcano and the ancient seafloor upon which it was constructed. (Modified from an illustration by D. Gillard and others)

it to the surface but solidifies underground. Thus Kilauea, like all shield volcanoes, grows from within as much as by layering of its upper surface. A good part of this internal growth takes place along the rift zones: each dike intrusion leaves a sheet of rock a few feet wide and perhaps several miles long. Many of these dike intrusions never break the surface at all. Thus, the rift zones are like stretch marks, gradually widening as they accommodate the mountain to the ever-growing body of rock within.

This style of growth has important consequences for the volcano's evolution. Kilauea's south flank—the sector of the mountain that is bordered to the north by the summit and the two rift zones—is gradually being wedged seaward. This movement is caused not only by the dike intrusions along the rift zones, but also by the gravitational instability of the flank itself. Although the parts of Kilauea's slopes that are above sea level are quite gently graded, the submarine portions are much steeper, on account of the rapid cooling of submarine eruptions brought about by contact with seawater. Thus, the south flank is gradually slipping into the Hawaiian Deep.

The south flank does not move in a single coherent block, however. Rather, it is cut by rows of normal faults that run parallel to the coast. Downward slipping of the seaward side of these faults produces characteristic cliffs called *pali*. Along much of Kilauea's south flank, so many lava flows have rolled over the pali that they have developed a smooth, rather gently curved outline, almost suitable for tobogganing (if there were any snow). Where the pali have been spared by the flows, however, they form spectacular, near-vertical cliffs.

There is believed to be another, even more important fault plane far below the surface of Kilauea's south flank. This is the plane marking where the volcanic edifice sits atop the original ocean floor. This plane is not quite horizontal, but slopes gently downward toward the center of the island.

Evidence that this "detachment plane" is an important earthquake fault came on November 29, 1975. At 3:36 in the morning almost everyone on the Big Island was awakened by a magnitude 5.7 earthquake. Among those especially disturbed were a party of Boy Scouts and others who were camping in a coconut grove at Halape Beach, directly south of Kilauea. Because of rockfalls that came down the 1000-foot slopes of Pu'u Kapukapu (the "hill of taboos"), some of the campers moved closer to the shore, where they eventually returned to sleep.

But the earthquake turned out to be a foreshock. Eighty-two minutes later, a much larger mainshock hit the island, damaging or destroying 85 homes, 11 commercial buildings, and 5 churches. It also caused numerous landslides, damaged some roadways, cut off electric power, and sent all the island's seismographs off scale. Later, data from seismographs at distant locations were used to calculate a magnitude of 7.2.

The campers at Halape Beach were badly thrown about by the earthquake. With rockfalls coming down Pu'u Kapukapu, they moved as close to the waterfront as possible. But within half a minute of the end of the shaking, they saw that sea level was rapidly rising. Although the campers ran back toward the hillside, many of them were knocked down by the inrushing wave. The wave receded, but was immediately followed by a second, larger wave that ran 100 yards inland. Many of the campers were thrown into a deep fissure, where they were churned about, along with horses, trees, rocks and debris from the Halape shelters. Several more waves washed over them as they struggled in the fissure. One man was killed there; another was washed out to sea and presumed drowned.

The tsunami rapidly spread up and down the coast, destroying piers, beachside houses, and vehicles. Boats were sunk both in Hilo Bay, on the east side of the Big Island, and at Keauhou and Kailua on the western (Kona) coast. The other Hawaiian Islands experienced only a very slight rise in sea level, but at Catalina Island, off the coast of southern California, the tsunami was large enough to cause minor damage to a dock.

As if earthquake and tsunami were not alarming enough, Pele herself was stirred into life. By 30 minutes after the mainshock, seismographs around the summit caldera were recording harmonic tremors. This ominous sign of rising magma was soon followed by an eruption: a 500-yard-long line of fire fountains opened up on the caldera floor. This eruption quickly abated, but it was followed by two smaller eruptions from the nearby wall of Halemaumau crater. The total eruption was quite minor in volume, but it was sufficient to illuminate the sky above the volcano with a ruddy glare, adding to the general sense of impending doom.

Surveys conducted later revealed that a part of the coastal area around Halape sank by as much as 11 feet during the earthquake, while regions to the east, such as Kaimu Bay, subsided by about 3 feet. Coconut groves that fringed the beaches now sat in seawater, and they soon died. The major surface faulting was along the normal fault system of the Hilina pali, a line of cliffs running parallel with the coast, about 3 miles inland. The coastal side of these faults sank by 4 to 5 feet, over a distance of about 16 miles. In addition, the coastal block moved seaward by about 25 feet.

Analysis of the seismic waves from the 1975 earthquake, and the distribution of the aftershocks, indicated that the principal movement was an outward sliding of the coastal block along the deep detachment plane, about 6 miles below the surface. How the normal faults of the Hilina pali fit in with this pattern of movement is not entirely clear. According to some scientists, the normal faults, once they reach a depth of a couple of kilometers, curve seaward and run horizontally until they reach the ocean floor several miles from the coastline. If this idea is correct, the downward slip on the Hilina pali faults during the 1975 earthquake was a secondary phenomenon, not closely related to the much deeper seaward motion on the detachment fault.

Paul Okubo, a geophysicist at the Hawaii Volcano Observatory, has published papers that are in accordance with this model. But when we spoke with him in 1997, he expressed skepticism about the earlier model and proposed instead that the Hilina pali faults dive straight down to the detachment plane. His reason for changing his mind was his elaborate analysis of seismic waves below Kilauea's south flank. By measuring the arrival times of many earthquakes at many locations on the volcano, and by analyzing the data in a manner akin to the way in which CAT scans of the human body are built up from X-ray data, Okubo was able to map the speed of sound in a huge three-dimensional space below the south flank. In his resulting images, there is a plane marked by a sudden change in the velocity of sound—corresponding to a sudden change in the composition of the rock—that runs at least 9 kilometers downward from the Hilina pali. This plane, he believes, is the Hilina pali fault system, running downward until it meets, or nearly meets, the detachment fault at the volcano's base.

If this model is correct, then a truly immense sector of the volcano, from its surface down to its base, moved about 25 feet seaward in the 1975 earthquake. The more central part of the volcano, north of the Hilina pali faults, did not move in 1975. But a similar earthquake in 1868 was probably caused by movement on a more northerly portion of the detachment plane, where it underlies the south flank of Mauna Loa. In that earthquake, both the south flank of Mauna Loa and a part of the overlying mass of Kilauea slid seaward. That earthquake was correspondingly larger than the 1975 earthquake—estimates of its magnitude range from 7.5 to 8.0. It killed 31 people and was followed by a tsunami that drowned 46 more.

"We're not just looking at a volcano, we're looking at a huge landslide system," Okubo told us. One is left wondering, not why the side of the volcano slides fitfully seaward, but what restrains it from the ultimate catastrophe—the sudden and total collapse of the volcano's entire flank into the ocean. As we will see later, such catastrophic collapses have in fact happened on many occasions over the course of geological time.

Adulthood: Mauna Loa

Twenty miles west of Kilauea, the broad mass of Mauna Loa rises 13,677 feet into the sky. Although the two volcanoes sit side by side, Mauna Loa is much older. It has been active for at least 500,000 years, and perhaps as much as 800,000 years. It is still in the main shield-building phase of its life, but this phase will probably come to an end within the next 100,000 years or so. Like Kilauea, Mauna Loa is erupting tholeiitic basalt. As mentioned earlier, however, there are subtle differences in the tholeiites erupted by the two volcanoes, indicating that they tap into separate magma sources.

Mauna Loa last erupted in 1984, before that in 1975, and before that in 1950. Thus, the impression has developed that Mauna Loa erupts significantly less often than Kilauea. But since 1832, the date of the first eruption recorded by Europeans, there have been a total of 39 documented eruptions, and quite possibly others that were not witnessed. Thus, the volcano has erupted, on average, at least once every four years during this period. The recent lulls in activity have been uncharacteristically long.

Geologist Jack Lockwood, who recently retired from a long-time position at the Hawaii Volcano Observatory, has made a thorough study of the lava flows currently exposed on the surface of Mauna Loa. Lockwood wanted to determine the ages of the flows, but potassium-argon dating and other such methods are generally applicable only to much older rocks (tens to hundreds of thousands of years of age). So Lockwood searched for plant material embedded in the flows and subjected it to carbon-14 dating. With this method he has determined the ages of 170 separate flows.

Lockwood's results suggest there has been a cyclical alternation of summit and rift-zone activity, similar to what has been inferred for Kilauea. For a few centuries, most eruptive activity is concentrated at the summit, which gradually builds in height. Then the summit collapses, forming a caldera, and a long period of rift-zone eruptions follows. Summit eruptions may continue during this period—in fact, most of the recent rift-zone eruptions have been preceded by small eruptions at the summit—but these eruptions are confined to the caldera, which slowly fills. Then another cycle of summit building begins. At Mauna Loa, the last two cycles have each lasted 1500 to 2000 years. Lockwood believes that we are currently reaching the end of a long period of rift-zone eruptions and that another summit-building phase may soon begin.

Because of its central location and great height, Mauna Loa is a threat to wide areas of the Big Island. In the 1950 eruption, which broke out during the night, lava flows emerged from vents on the southwest rift zone, raced down the relatively steep western slopes of the volcano, cut the coast road, and destroyed several small communities, before running into the ocean. Only speedy evacuation saved the

lives of the inhabitants. Within a three-week period, Mauna Loa erupted more than half a billion cubic yards of magma—an amount corresponding to several years' worth of Kilauea's output.

In the 1984 eruption, the lava flows went the other way, toward the
east coast. Although the eruption began at the summit, it quickly shifted to vents on the northeast rift zone. The slopes are gentler on the east
side of the volcano, but in time lava tubes formed and helped carry lava
toward the city of Hilo. For several weeks, the residents of the city had
to put up with the unnerving sight of a line of fire advancing on them
from the west. The flows stopped 4 miles short of the city.

"Everyone's very curious when Mauna Loa is going to erupt
again," Christina Heliker told us. But she also said that she and her colleagues are unable to monitor Mauna Loa as closely as they would like.
Visiting Mokuaweoweo, the summit caldera, entails either an all-day
hike or an expensive helicopter ride. Expense has also prevented the
installation of permanent, continuously operating GPS stations on the
mountain. To monitor deformation of the mountain, it is necessary to
bring equipment up and take measurements in person, so this is generally done no more than once or twice a year. Seismographs, however,
are in place. Because the last two eruptions of Mauna Loa were preceded by about a year of shallow seismic activity under the summit,
Heliker hopes that the seismographs will give good advance notice of
the next eruption too, allowing the scientific staff to transfer more of
their resources to monitoring the mountain. But an eruption with just
a few hours' seismic warning is always a possibility.

As with Kilauea, sectors of Mauna Loa are being wedged seaward
by expansion along the rift zones. Thus, landslides are a potential hazard. Several times in the volcano's history, enormous chunks of the volcano's west flank—hundreds of square miles at a time—have fallen cat-
astrophically into the ocean. The most recent of these landslides—the
so-called Alika phase 2 slide—took place about 100,000 years ago. The
rubble from the slide rushed into the Hawaiian Deep with such velocity that it was carried 50 miles across the ocean floor and up the opposite slope of the Hawaiian Arch. This slide, along with many other
prodigious Hawaiian landslides, has been mapped by Jim Moore of the
USGS at Menlo Park and several colleagues, who used acoustic mapping, submersible exploration, and bottom dredging to outline the
remains of the landslides on the ocean bottom and to figure out which
volcano each slide came from.

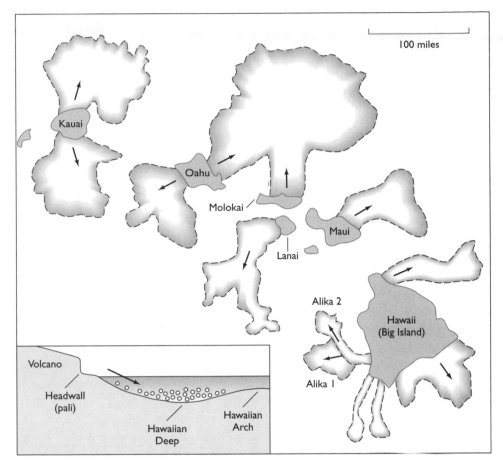

Map of giant submarine landslides around the Hawaiian Islands. The inset shows how the slides originated at the pali and filled the Hawaiian Deep with rubble (not to scale). The Alika 2 slide was the probable cause of a giant tsunami that inundated Lanai, Maui, and Molokai about 100,000 years ago. (Adapted from an illustration by J. Moore and others, USGS)

In the early 1980s, Jim Moore and George Moore (no relation) made a discovery that illuminated the potentially deadly power of these landslides. Several geologists working on the islands to the north of the Big Island had noticed limestone boulders and pieces of coral on some of the seaward slopes, as much as 1000 feet above the present shorelines. How had these rocks, which must have been formed below the ocean surface, found their way up the slopes of Lanai, Molokai, and Maui? Various opinions were offered. Perhaps the sea level had once been

1000 feet higher than it is now. Perhaps native Hawaiians had carried the rocks up—for some ritual purpose, for the exercise, or to perplex future geologists.

The two Moores disproved these hypotheses, and they presented strong evidence for another: the rocks were carried up by a tsunami wave. The main evidence was similar to the evidence used to identify tsunami deposits in the Pacific Northwest; namely, that the deposits become progressively thinner and finer-grained inland and uphill. In addition, the deposits are particularly thick where the backflow of the wave was hindered by the topography—such as in Kaluakapo crater on Lanai, which acted as a catch basin.

Radiometric dating of coral fragments suggests that the tsunami deposits are about 100,000 years old. Thus, a reasonable candidate for the cause of the tsunami is the Alika phase 2 slide from Mauna Loa. Since the time of the tsunami, the islands affected by it have sunk 500 feet or more (by a mechanism we'll describe later); therefore, the tsunami's actual run-up height was probably closer to 2000 feet than to 1000. When one considers that the greatest tsunami to hit the Hawaiian Islands in modern times had a maximum run-up height of only 66 feet (this was the 1946 tsunami—caused by an earthquake in Alaska—that killed 159 people), the consequences of a *2000-foot* run-up are scarcely imaginable. Very likely, the majority of the residents of the Hawaiian Islands would be killed. Luckily, the giant sector collapses, as they are called, are very few and far between: one occurs only every few hundred thousand years, on average.

A Volcano in Its Prime: Mauna Kea

Skipping over Hualalai, a volcano to the northwest of Mauna Loa on the Kona coast, we come to Mauna Kea, the predecessor to Kilauea in the eastern row of volcanoes. Mauna Kea is about 900,000 years old. At 13,796 feet, it is the highest volcano in the Hawaiian Islands. It represents that brief cusp in volcanic evolution—after long ages of eruptive fury have raised the edifice to its highest point and before long ages of decay have torn it down.

Yet that process of decay, as with human life, begins already at the moment of conception. Several forces conspire, right from the beginning, to undo a volcano's upward striving. We have already mentioned two: the sagging of the underlying lithosphere, which directly counteracts a portion of the volcano's growth, and the slumps and landslides that jettison much of the volcano's hard-won bulk into the Hawaiian Deep. But there are other factors, too. Erosion brings down all mountains eventually. The underlying lithosphere, temporarily heated and swollen during its passage over the hot spot, later begins to cool and contract, drawing the overlying volcanoes downward. Finally, there is another, even slower process: the cooling and sinking of the entire Pacific Plate as it drifts ever farther from its origins at the mid-ocean ridge.

As long as a volcano is in its shield-building phase, its growth easily outpaces these downward forces. But Mauna Kea has nearly exhausted its inner fire. Its last eruption was about 3000 years ago. Almost certainly, it will erupt again—perhaps many times—but never with the prodigious flows that raised it from the ocean bottom to its present height in less than a million years.

Like Kilauea and Mauna Loa, Mauna Kea erupted tholeiitic basalts for most of its shield-building years. But about 200,000 years ago, the lava flows changed composition, to a form of basalt containing slightly more silica (about 52 percent versus 49 percent for tholeiite) and also enriched in the alkali metals, sodium and potassium. This difference in composition was enough to make the late-stage *alkalic* lava flows more viscous than the earlier, tholeiitic flows. Therefore, in its last stages of growth, the volcano became more rugged and more steeply sloped than previously. Seen from Hilo, the difference between the outlines of Mauna Loa and Mauna Kea is striking. Mauna Loa is as broad as an elephant's back; Mauna Kea is that same elephant, laboring under an alkalic howdah.

The change from tholeiitic to alkalic lavas late in a volcano's career is not unusual for the Hawaiian Islands. In fact, among the old volcanoes northwest of Mauna Kea, only Lanai and Oahu's Koolau volcano lack an alkalic cap. Alkalic lavas have also been dredged up from the summits of the sunken seamounts to the north. Probably no Hawaiian volcano erupts more than about 1 percent of its total volume in the form of alkalic lava, but because this lava lies atop all the earlier flows, it is hard to miss.

According to Ed Stolper, a Caltech geologist who is currently studying Hawaii's lava flows, the switch from tholeiitic to alkalic lava occurs at a critical moment in the passage of the volcano across the underlying mantle plume. During the tholeiitic phase, the volcano is fed by magma that rises more or less directly from the central portion of the plume. This magma, as described earlier, is thought to come from the deep mantle, with entrainment of some "primordial" material from the middle mantle. Eventually the volcano moves beyond the reach of the central part of the plume. In Stolper's model, however, the rising plume partially melts the rock in a surrounding region of the asthenosphere—the very upper part of the mantle immediately under the crust. Thus, as the volcano moves from above the central zone of the hot spot and enters this "halo" region, it is fed for a while by magma from the upper mantle. It is this magma, Stolper believes, that is the source of the terminal alkalic flows.

Of course, the site of a Hawaiian volcano must actually cross the halo region twice: once before it reaches the hot spot, and once as it leaves it. Thus, if Stolper's theory is correct, one might expect there also to be an initial alkalic phase, prior to the main tholeiitic sequence. For most of the mature volcanoes, the flows deriving from this initial phase would be buried tens of thousands of feet beneath the present surface. The one site where it might be reasonable to find these early-phase alkalic flows is on Loihi. In fact, Alexander Malahoff and others have found some alkalic flows exposed near Loihi's summit. According to Malahoff, however, the findings don't fit in too well with Stolper's ideas, because the alkalic flows on Loihi are not the earliest flows; there is no suggestion in his data that alkalic eruptions preceded the tholeiitic eruptions.

One way to resolve this issue would be to drill a hole straight down through a mature volcano, with the idea of recovering a core that illustrated much or all of its life history. Exactly such a project has been undertaken by a consortium of scientists led by Donald DePaolo of the University of California, D. Thomas of the University of Hawaii, and Stolper. The drilling team has already completed a test hole, which was located near Hilo. The hole passed through a portion of the eastern flank of Mauna Loa and then entered the buried portion of Mauna Kea's edifice. The drilling was terminated at a depth of 3000 feet, where the rocks had an age of about 400,000 years.

The main hole, which is planned to be completed in about six years, will go down about 15,000 feet. At this depth, the scientists expect to find rock about 650,000 years old—quite likely corresponding to an even earlier stage of volcanic development than the stage at which Loihi currently finds itself. Of course, the project has many more goals than just resolving the question of an early alkalic phase. In fact, the drill core—as well as the thermal, magnetic, and other measurements that will be taken within the hole—promises to open a wide new vista onto the evolution of hot-spot volcanoes.

Senescent Volcanoes:
Northwest from the Big Island

The volcanoes to the northwest of Mauna Kea are in various stages of disintegration and sinkage. Landslides have been a big part of the disintegration. All of the Hawaiian Islands show traces of such slides, but the biggest was a slide that removed the northeast flank of Koolau volcano, on Oahu, at some unknown time in the past. According to estimates by Jim Moore, more than 1000 cubic miles of the volcano crashed into the ocean. (Compare that with the 1980 Mount St. Helens landslide—one of the largest in modern history—which had a volume of less than *one* cubic mile.) The rocky debris, including blocks a mile or more across, now lies scattered over 9000 square miles of ocean floor. The headwall of the avalanche is marked by the steep cliffs of the Nuuanu pali, 3 miles inland from Oahu's east coast. In 1795, when Kamehameha the Great conquered Oahu, his troops threw the defenders over these cliffs; more than 500 skulls have been found at the bottom.

Erosion also plays a major role in the decline of Hawaiian volcanoes. The best views of erosion at work can be had on Kauai, the northwesternmost of the major islands. Kauai is the remnant of a single enormous shield volcano, Waialeale, that is estimated to be about 6 million years old. Thus, there has been plenty of time for running water to attack the mountain. And there has been plenty of water, too: the island's 5158-foot summit is the wettest spot on "dry" land, receiving an average of 486 inches (40 feet) of rain per year. So Kauai is cut by gorges thousands of feet deep: Hanalei Valley to the north, Hanapepe Valley to the south, and Waimea Canyon (the "Grand Canyon of the Pacific") to the west. In the walls of these canyons, the endlessly layered lava flows and ashfall deposits tell of hundreds of thousands of years of volcanic evolution and thus, to a degree, make drilling projects superfluous.

Besides classical erosion, which is caused by abrasion from water- and airborne particles, the Hawaiian volcanoes are also under chemical attack. Naturally acidic rain eats away at the lava flows, slowly dissolving them. Those crystal clear streams that tumble over Hawaii's famous waterfalls are actually carrying lava—or dissolved minerals from the lava—to the ocean. Because rainwater lingers in shady spots, it is the clefts in the pali and valley walls that are eaten away the fastest. Thus, these cliffs gradually develop the fluted, cathedral-like architecture— deep recesses alternating with tall buttresses—for which the Hawaiian Islands are famous.

The volcanoes don't go under without a few last feeble efforts to rebuild their edifices. For reasons that are unclear, many of the volca- noes erupt in fits and starts over a period of a few million years after the main tholeiitic and alkalic phases are over. Such eruptions have taken place on Kauai, for example, as recently as 500,000 years ago. The vol- umes erupted during this rejuvenation phase are trivial compared with what went before, but because they are the most recent eruptions, they provide much of the currently visible scenery. Well-known landmarks such as Oahu's Diamond Head are the product of rejuvenation-phase volcanism.

There are quite a few puzzling aspects of these late-stage eruptions. For one thing, the eruptions don't follow the ancient conduits but crop up at what seem to be quite arbitrary locations. In addition, the lava flows produced by these eruptions, although mostly belonging to the class of alkalic lavas, contain bits and pieces of unusual materials—rocks whose exotic status is hinted at by their names: spinel lherzolite, rare garnet peridotite, and so forth. The origin of these *xenoliths* has been disputed in many a learned paper. They probably come from the upper mantle, but why they are seen particularly in the late-stage eruptions is unclear. Just as a patient with pneumonia often has to cough up a series of sputum samples before the really interesting bacteria come forth, so the volcano may have to clear out most of its supply of magma before the long-accumulating detritus of exotic rocks is brought to the surface.

Besides being worn away by landslides and erosion, the Hawaiian volcanoes sink en bloc, because of the sagging of the underlying crust (caused by the weight of the volcano) and later because of the cooling and contraction of the lithosphere, as described earlier. As revealed by tide gauges along the east coast of the Big Island, Mauna Kea is sinking at a rate of 2–3 millimeters per year.

The history of an island's sinkage is revealed by submarine land- marks. As mentioned earlier, lava flows on land form gentler slopes than those formed underwater. Thus, until the end of a volcano's shield-building phase, there is a change in the angle of slope—a "slope break"—at or near the current shoreline. After the volcano ceases to build its edifice, however, continued sinkage takes the slope break beneath the waves. According to Jim Moore, Mauna Kea's slope break is now 1200 feet below sea level. The slope break for Haleakala, the largest volcano on Maui, is now about 6000 feet underwater. Even if the island of Maui were somehow protected from all further landslides and erosion, the inexorable process of sinkage would bring the summit of Haleakala down to the waterline less than 3 million years from now.

The Wreath on the Coffin: Coral

Yet just when the summits of the Hawaiian volcanoes are preparing to plunge forever beneath the waves, a host of little animals comes to the rescue. Coral reefs are found already on the younger volcanoes, but they do not take major hold until late in the process of sinkage, when the underlying crust has sagged about as much as it is going to, and only the cooling-contractional process is still operative. Under these conditions, coral grows upward as fast as the volcano descends, always keeping its living members near the water line. (This process, by the way, was first understood by Charles Darwin, who visited the coral Cocos Islands during his famous voyage on the *Beagle*.) Thus the Leeward Islands— the line of tiny coral islands, shoals, and reefs than run northwestward from the main islands of Hawaii—mark the peaks of the chain of descending volcanoes, each bearing a taller crown of coral than the next.

But the coral reefs only postpone the inevitable. As each volcano drifts northwestward, the cooler waters become less favorable to coral growth. At long last, the coral-crowned volcano slides beneath the waves, like Ahab's *Pequod,* which "would not sink to hell till she had dragged a living part of heaven along with her, and helmeted herself with it."

Many other oceanic islands owe their existence to hot-spot volcanism. We already mentioned Reunion in the Indian Ocean and Iceland in the North Atlantic. Lonely Tristan da Cunha is Iceland's counterpart in the South Atlantic: it also sits astride the Mid-Atlantic Ridge. An eruption in 1961 forced its 200 inhabitants to seek temporary refuge in England. The Canary Islands, off Africa's northwest coast, are the site of another hot spot. Over the course of their geological history, these seven islands—which range up to 12,000 feet above sea level—have been the sites of numerous sector collapses, some of which were larger than the biggest landslides in Hawaii. The Galapagos Islands, in the Pacific off Ecuador, represent the activity of yet another hot spot; this one may have began its career by creating a flood-basalt province in the Caribbean.

The story of oceanic hot spots illustrates, as well as any topic in this book, the inexorability of geological processes and the inevitability of further change. Maui will sink beneath the waves, and Loihi will rise from them; of this, we can be as certain as we can be of anything in the world's future. And with equal certainty, these processes will be accompanied by great eruptions, earthquakes, landslides, and tsunamis—all capable of causing enormous human catastrophes. So what can we or should we do about the threats? These questions are the theme of our final chapter.

13

THE YEARS OF LIVING DANGEROUSLY: HOW DO WE SURVIVE ON AN EARTH IN TURMOIL?

Over the past century, scientists have made extraordinary progress in understanding the causes of earthquakes and volcanic eruptions. We now know that these sudden catastrophic events are caused by imperceptibly slow, seemingly harmless motions within the Earth that nevertheless gradually load the Earth's crust toward failure or bring the Earth's inner heat dangerously close to the surface. By mapping the sites where these processes occur—the plate boundaries, subduction zones, hot spots, and so forth—and by analyzing the record of previous eruptions or fault ruptures at these locations, we can draw conclusions about what may be expected in the future. This accomplishment offers a notable benefit to humankind: the possibility of averting the harmful consequences of these natural disasters through advance preparation. The value of this scientific understanding, and of the consequent preparations, seems particularly great in locations such as the coast of the Pacific Northwest where, aside from obscure Indian legends, human records of past catastrophes are lacking.

Yet, great as these scientific accomplishments have been, they are also painfully inadequate. First, a small fraction of the world's earthquakes and volcanic eruptions don't fit clearly into the framework of knowledge that has been established. We know a lot about why earthquakes occur in California but little about why they occur in Missouri or Massachusetts. We know a lot about why volcanoes erupt in Cascadia or Hawaii but little about why they erupt at Mammoth Lakes. Plate tectonics, as we presently understand the subject, has not answered every question about why certain regions of the world are subject to geological catastrophes.

Second, the knowledge that we have gained so far gives us only a very general power to predict future events. We can make some estimate of the probability of an earthquake or volcanic eruption in a given region over a given period of time, but we cannot pin these events down to specific places and dates. Even where circumstances have seemed unusually favorable for making such predictions, such as on the San Andreas fault at Parkfield, the results of these predictions have been disappointing so far.

Of course, there is a major difference in this respect between earthquakes and volcanic eruptions. Volcanic eruptions are preceded by the movement of magma from deep reservoirs toward the surface. This movement usually gives warning signals: earthquake swarms, deformation of the volcanic edifice, and gas emissions. No volcano, properly monitored, is likely to erupt without any warning. Thus, there is every reason to invest in such monitoring techniques and to research ways to improve them.

Earthquakes, on the other hand, are caused by the sudden failure of rock that has been stressed to its breaking point. There is no certainty that there will be any detectable warnings signs in the days or weeks before the impending failure. That's not to say that there are never such signs. Some major earthquakes have been preceded by increases or decreases in the ongoing pattern of seismicity, for example. On very rare occasions, such as in the 1975 earthquake near Haicheng, China, the timely detection and interpretation of these trends have saved many lives. It's certainly worthwhile continuing to look for precursory events. But it is likely that detectable precursors will be few at best, and they may occur only in certain locations. This is why we have devoted little space in this book to research on earthquake precursors.

An ability to provide short-term warning of earthquakes might have prevented the loss of perhaps a million lives this past century. Nonetheless, for many purposes, precise short-term warnings are useless: one cannot strengthen a city's buildings or bridges in a matter of weeks or months, let alone in minutes or hours. Most earthquake mitigation requires long-term planning, and for such planning the general location, magnitude, and probability of future earthquakes are all that is required. The knowledge we do have in some places and are developing elsewhere—knowledge about the likely location of earthquakes and the general probability of their occurrence over medium-to-long time spans—could help prevent most earthquake casualties, if only it could be translated into effective measures to mitigate the earthquakes' damaging effects.

Earthquake mitigation, like mitigation of natural disasters in general, is an enormously challenging enterprise. Acquiring and interpreting the necessary scientific information—the main topic of this book—are just the beginning. Between this knowledge and the appropriate response come heavy doses of politics, economics, human psychology, and even chance, all of which speed or slow the process in unpredictable ways.

Mitigation—A Case History

Let's take a look at one particular mitigation effort: the strengthening of unreinforced masonry buildings (or URMs) in the city of Los Angeles. The history of this effort, which was begun 25 years ago and is still incomplete, illustrates some of the strategies, pitfalls, and successes of earthquake mitigation.

The story begins in the aftermath of the 1971 San Fernando earthquake (magnitude 6.7), which caused 58 deaths and 2500 injuries, as well as about half a billion dollars' worth of property damage. Analysis of the damage led quickly to a number of state laws, city ordinances, and code changes that involved retrofitting various systems within existing buildings. For example, many of the 109 fires that broke out after the earthquake were caused by falling gas water heaters, and an ordinance was passed requiring the heaters to be fastened to wall studs. More than 100 elevator cars were damaged or destroyed, in most cases because the elevators' massive counterweights jumped from their guide rails and collided with the cars. An ordinance was passed that mandated modifications of the elevators to prevent a recurrence. And in schools near the epicenter, many hundreds of heavy lighting fixtures fell from the ceilings. (Luckily, the classrooms were empty at the time.) In response, a state law was passed to require improved anchoring of the fixtures.

The problem of the basic structure of URMs, however, was much more recalcitrant than any of those just mentioned. The susceptibility of such buildings to catastrophic collapse had been known for more than a century. Typically, one or more walls collapse outward, leaving the floor and roof systems unsupported. These immediately pancake downward, crushing the occupants before they have a chance to leave the building. A wood-frame or steel-reinforced building, in contrast, has enough ductility to ride out most earthquakes, especially if it is secured to its foundations. If it does collapse, it will do so only after a considerable period of shaking that may give occupants time to escape, and the collapse typically leaves large voids where trapped occupants may survive.

After the 1886 earthquake in Charleston, South Carolina, many brick buildings in that city were reinforced with wall anchors that tied the outer walls to the floor and roof systems in an attempt to prevent the walls from blowing out. The effectiveness of the Charleston devices has yet to be tested, as that city has not experienced a major earthquake since 1886.

A much more ambitious retrofitting effort was begun in the city of Long Beach (south of Los Angeles), after a 1933 earthquake (magnitude 6.3) that took 115 lives and damaged or destroyed more than 1000 URMs. (The earthquake was caused by a rupture of the Newport-Inglewood fault, a strike-slip fault that accommodates part of the relative motion between the Pacific and North American plates.) The Long Beach ordinance, besides setting seismic standards for new construction, also required old URMs to be brought up to the standards of modern buildings. This requirement proved so expensive that the owners of most of the affected buildings—including a number of buildings of historic significance—chose to demolish rather than strengthen them.

The Long Beach experience was a strong disincentive to similar action in Los Angeles. It was estimated that of the 8000 or so URMs in Los Angeles, more than 1000 were apartment buildings or hotels, containing a total of about 45,000 dwelling units. Many of these units were low-income apartments shared by several families—often immigrants. The prospect that 100,000 to 200,000 poor people might be evicted from their homes, in a period of severe housing shortage, was a daunting one. The demolition of the remaining, nonresidential buildings would also pose serious economic or logistical problems. Many of the city's firehouses, for example, were in this category.

Shortly after the San Fernando earthquake, councilman Tom Bradley sponsored a resolution that authorized the city's Building and Safety Department to look into the problem of URMs. But little if anything was done for several years. In 1975, however, help came from an unexpected quarter. Another councilman, Art Snyder, wanted to get rid of the pornographic movie theaters that had sprung up in his district. He approached the chief of the Conservation Bureau, the bureau within the Department of Buildings and Safety that deals with existing buildings. A quick drive-around revealed that the porno theaters were housed in old brick buildings. So Earl Schwartz, a structural engineer in the bureau who later became chief of the new Earthquake Safety Division, drafted a preliminary ordinance that required owners to upgrade their buildings to present-day codes—not just the seismic codes but the electrical, mechanical, and plumbing codes, too. This ordinance, it was hoped, would be so burdensome that it would put the porno theaters out of business.

To legitimize this piece of chicanery, however, the ordinance had to be broadened to include *all* movie theaters, and eventually to include all URMs whatever their function. This led to massive resistance on the part of property owners. Public hearings on the matter, held in 1976 and 1977, became battlegrounds in which apartment owners and their bused-in tenants flooded officials with dire warnings about mass evictions and homelessness. To placate them, a proposal was put forward that owners of unreinforced masonry apartment buildings could choose, instead of strengthening them, to post notices that the buildings were hazardous in an earthquake. The owners strongly resisted this proposal, too. Eventually, the entire ordinance was sent back to the Building and Safety Department with the instruction to further study the issue and come up with revised proposals within two years. In the process, the ordinance's original purpose—to harass the owners of porno theaters— was quietly forgotten. During these years of indecision, disastrous earth- quakes in Guatemala (22,000 dead) and China (at least 250,000 dead) accentuated the importance of seismic retrofitting for its own sake.

Schwartz formed a team of about 10 inspectors who surveyed the city's inventory of URMs and set out to develop standards for seismic strengthening. With input from a citizens' committee, Schwartz and his colleagues realized that strengthening of the URMs to present-day standards was not economically feasible and that a level of earthquake resistance 60–70 percent of that mandated by new-construction codes was all that could be hoped for. In addition, they decided to make allowances for whatever seismic resistance the existing buildings might possess. To that end, they acquired a vacant city-owned apartment building and subjected it to all kinds of mechanical abuse, with the aim of developing techniques to assess shear resistance, strength of mortar, and so forth.

It was late 1980 before the revised ordinance was reviewed by the city council. As it happened, the preceding 12 months had seen a notable sequence of geological disasters. The Imperial fault had rup- tured in late 1979, damaging or destroying about 2000 buildings in the Imperial Valley (see Chapter 5). Mount St. Helens erupted in May 1980, and the Long Valley caldera threatened to erupt soon thereafter (see Chapters 2 and 7). Earthquakes played a well-publicized role in both those events. And in late 1980, earthquakes in Algeria and southern Italy claimed a total of about 8000 lives; as was readily apparent to television viewers, the collapse of unreinforced masonry buildings was a major cause of the fatalities. These disasters helped to create a climate at least temporarily favorable to mitigation efforts.

Another factor was the recent election of a new member of the city council, Hal Bernson. As is customary for new council members, Bernson was appointed to the chair of a politically unimportant com- mittee, in this case the Building and Safety Committee. Conservative by instinct, Bernson was not well disposed to the notion of a mandatory retrofitting ordinance. He did have considerable personal experience of earthquakes, however: as a child growing up in the Bakersfield area, he had lived through a magnitude 7.5 earthquake—the strongest to strike California since 1906. And his district in the San Fernando Valley was among those hardest hit by the 1971 earthquake.

Earl Schwartz did everything he could to sell Bernson on mitigation. The two men went down to El Centro, in the Imperial Valley, to view the effects of the 1979 earthquake. Schwartz was able to show Bernson how strengthening measures, where they had been carried out, had reduced or prevented damage. Bernson was convinced; he became the sponsor of the ordinance, and he has been a leading voice for earthquake mitigation ever since.

The revised ordinance not only required less extensive strengthening than the original draft, but it also responded to the owners' economic arguments by setting a more relaxed timetable for the program. Owners needed to do nothing until notified, and notification might be delayed for several years, depending on the type of building. And if owners, once notified, promptly completed the first phase of the retrofit—the relatively inexpensive installation of wall anchors—they could delay the more elaborate work, involving the installation of shear-resistant elements, for several more years. It was estimated that it would take 15 years to complete the retrofitting of all 8000 URMs in the city.

The ordinance was passed in a preliminary vote of the council in December 1980. Before the vote could be made final, however, the apartment owners made one last stand. The owners' association printed a flyer entitled "Notice of Intended Eviction or Rent Raise," which told their tenants, in English and Spanish: "If you do not want to be EVICTED, or if you do not want your RENT DOUBLED, meet on the steps in front of the Los Angeles City Hall on Tuesday December 23rd, 1980, at 9:00 a.m., and demand that the earthquake law be defeated."

The final vote was postponed until January 7, on which day apartment residents showed up in droves. Among them, as reported by the *Los Angeles Times,* was 14-year-old Ruth Gurrola, who lived in an old brick building in the downtown area. "If they want to kick us out," she pleaded to the council, "I want a new apartment. I don't want to go on Skid Row, as a wino. We want to stay there, have a park, get an education."

In the end, 11 council members were stony-hearted enough to vote for the ordinance, and only 3 opposed it. One factor in the decision was Bernson's prediction that private, state, or federal funding would be forthcoming to ease the financial burden imposed by the ordinance. In fact, this funding never materialized; when federal dollars finally started flowing, 15 years after the ordinance was passed, they were to finance the *demolition* of URMs—the very outcome the council was trying to avoid in 1981.

Things did not move very fast after the ordinance was passed. For one thing, the apartment owners sought a court injunction to prevent enforcement of the ordinance, but it was denied. In addition, it was another two years before the council gave permanent budget authority to hire staff for the project. By 1985, the project was about two years behind schedule—only the owners of the most critical buildings had received orders to carry out the retrofit, and few of these had actually begun work. But in September of that year, an earthquake killed more than 8000 people in Mexico City. Again, the poor performance of URMs (and precast reinforced concrete buildings) was apparent to everyone who witnessed the rescue efforts on television. Under pressure from Bernson, the ordinance was revised to speed up the program, and funding was provided to allow more than 50 city employees—engineers, inspectors, and support staff—to work on enforcing it. Gradually, masonry buildings were fitted with wall anchors. The retrofitted buildings are easy to recognize from the street because of the rows of square metal plates that cap the ends of the anchor bolts. The plates have become a kind of regional architectural detail, like the protruding roof beams of houses in New Mexico or the wrought-iron balustrades of New Orleans.

Seismological research also propelled the project forward. Paleoseismic evidence for an approximately 130-year recurrence interval for ruptures of the San Andreas fault (see Chapter 4) helped convince politicians such as Bernson that another great earthquake on the San Andreas fault was inevitable. And in 1983, geologists first correctly interpreted a damaging earthquake as arising from slip on a blind thrust fault. (It was an earthquake near Coalinga, in California's Great Valley.) Only four years later, an earthquake occurred on a blind thrust fault in the Los Angeles area. (It was the small Whittier Narrows earthquake, which killed eight people and caused $200 million in damage.) Thus, the hazardous nature of the local terrain was becoming increasingly apparent. Furthermore, the tectonic processes that are inexorably squeezing the Los Angeles basin began to be understood and measured. This effort led to the realization that the basin was "behind schedule" in terms of its earthquake budget, so that the frequency of damaging earthquakes was likely to increase. Increased contacts among scientists, city planners, and the general public ensured that this scientific knowledge impacted on the decision makers. This interaction was substantial-ly promoted by the new Southern California Earthquake Center, a consortium of universities and governmental agencies funded by the National Science Foundation and the USGS.

Thus, by the time of the 1994 Northridge earthquake, the URM retrofit program was about 90 percent complete: about 6000 buildings had been strengthened, about 600 had not yet been strengthened, and the remainder had been demolished. In the aftermath of the earth-quake, a task force was set up to evaluate the effect of the earthquake on buildings, and Larry Brugger, who was director of the Earthquake Safety Division at the time, headed a subcommittee to investigate the performance of the URMs.

According to Brugger, as well as other experts who participated in the task force, the retrofitting was an outstanding success. Not a single person died in an unreinforced masonry building. Many of the non-retrofitted buildings sustained major damage, including wall blowouts, even though most of these buildings were situated in the eastern parts of the city, far from the epicenter. The lack of casualties in these build-ings was mainly a result of the fact that the majority of them were commercial properties; because the earthquake struck before dawn, most of the buildings were empty.

Of the retrofitted buildings, many did suffer significant damage, such as cracking of the masonry columns between windows, but less than half a dozen buildings suffered even partial collapse. In many cases, according to Brugger, the damage that did occur could be traced to inadequate implementation of the retrofitting standards: the presence of a bathroom, for example, sometimes prevented the installation of anchors at one point along a wall, and this portion of the wall cracked or collapsed. Or the presence of veneers over a wall might have led to erroneous estimates of the thickness of the masonry and thus to inade-quate bracing.

Most of Los Angeles' stock of URMs experienced only moderate shaking in the Northridge earthquake. How they would fare in a mag-nitude 7.5 earthquake centered near downtown (see Chapter 6) remains unknown. Still, on the evidence so far, the URM program must be considered a success: in all probability, it saved hundreds of lives in the Northridge earthquake alone. The city is now attempting to wrap up the program by strengthening or demolishing the 400 or so URMs that have not yet been retrofitted.

The URM ordinance was followed by a series of similar ordinances that dealt with other kinds of potentially hazardous buildings: buildings constructed on hillsides, buildings constructed of prefabricated concrete wall units ("tilt-ups"), buildings with open first floors ("soft-stories"), and so on. Most of these ordinances have lacked enforcement provisions, however, and compliance is generally low. The political and economic climate at present is not favorable to mandatory retrofitting, even though the Northridge earthquake showed that URMs are not the only vulnerable structures in Los Angeles. Whether this situation represents a reasonable strategic shift or a fatal failure of will remains to be seen.

Costs and Benefits

It must be disheartening to planners when the very people an ordinance is intended to benefit—the residents of unsafe buildings—are vociferous in opposing it. How different this situation is from, say, the field of commercial aviation, where passengers typically clamor for safety improvements regardless of cost or inconvenience. And even the resistance of the buildings' owners seems unreasonable: although the retrofitting may not save them personally from injury or death, it commonly increases the value of the building by an amount comparable to or greater than the cost of retrofitting, and it often makes a building insurable that previously had not been. Why, then, the opposition?

Several psychological and economic factors seem to be at play here. First, the people who live in unsafe buildings, especially the URMs, are typically poor people who live from day to day and cannot afford the luxury of long-range planning. A small increase in rent, or even a temporary eviction, may push such people into the abyss of homelessness. Even more affluent people, such as many of those living in "soft-story" apartment buildings, may have little disposable income to spend on averting any of a whole range of potential calamities.

The state of California's seismic safety codes are explicitly designed to reduce seismic risks to an "acceptable" level, but acceptable to whom? To the people who are exposed to the risks, one would imagine. But if so, have not these people, by their inaction, already made an implicit decision to accept them? Have they not made a conscious or unconscious assessment of all the circumstances of their lives and decided that, in these circumstances, the best thing for them to do is nothing? According to this kind of thinking, seismic safety codes should be voluntary; residents should be provided with information about the risks and how best to abate them and then left to do what they want.

This philosophy, of course, is part of a broad conservative ideology that is quite powerful in American society. Thus, it should not have been surprising when, during the Los Angeles City Council hearings on the URM ordinance in 1980, the prominent tax-reform crusader Howard Jarvis showed up to attack the ordinance. Jarvis's philosophy is still influential today, and it is part of the reason that most of the current ordinances are voluntary or lack enforcement provisions.

Several lines of argument can be raised against this laissez-faire attitude. First, one can point out the complexity of modern society and technology and the resulting impossibility for individual citizens to make well-informed, independent decisions on every matter that might affect them. Should not a democratic government, by virtue of its unique ability to obtain expert advice in almost any matter of interest, feel entitled to pressure citizens into doing the right thing?

There is also the "no man is an island" argument: the idea that society as a whole is affected by losses sustained by individuals. This argument can be quantified in terms of cost-benefit or cost-effectiveness analyses. In the case of earthquakes and volcanic eruptions, societal losses might include the cost of search-and-rescue efforts, medical treatment, insurance losses, lost productivity, care of orphaned children, emergency housing, relief grants, low-interest loans, and so on. Rakish Sarin of the UCLA School of Management carried out one such study of the URM ordinance shortly after it was enacted. He concluded that, in general, societal interests would be best served by lower standards than those actually enacted. Residential URMs, he said, should be retrofitted to a lower level of seismic resistance than the ordinance actually required, and commercial URMs should not be subject to mandatory retrofitting at all. Because of the lack of financial incentives to the owners, Sarin predicted that they would put up continued resistance to the program, leading to unacceptable enforcement costs. In fact, however, owner compliance turned out to be quite good, especially after the Mexico City earthquake of 1985. Either Sarin underestimated the economic benefits of retrofitting to the owners, or imponderable factors such as sheer altruism or the dramatic footage of the Mexico City earthquake finally swayed them toward compliance.

Finally, there is the argument that, in writing mandatory ordinances, the government is mediating conflicts of interests among citizens—a well-accepted governmental function. Such conflicts of interest are obvious in the case of apartment buildings or commercial buildings. Even owner-occupied single-family homes, however, can be covered by this argument: government often intervenes as a representative of children's interests—smoke-detector and lead-paint ordinances are prominent examples.

No one carried out a detailed analysis, after the Northridge earthquake, to determine whether the URM ordinance had proved itself worthwhile, either in a cost-effectiveness sense or in a broader human context. In particular, there has been no detailed study of the effect of the ordinance on housing availability. There were reports in 1985, as the city of Los Angeles accelerated its enforcement of the ordinance, that considerable numbers of people—perhaps a few thousand—did have to leave their homes. Either they were evicted without relocation assistance or rents were increased beyond their means. (Landlords were allowed to raise rents if the seismic upgrading was accompanied by other improvements in the property.) At this point, it seems impossible to establish how many people were affected or what became of them. With that caveat, however, it seems that ordinance has on balance been very beneficial. As the decades go by, of course, the question of whether the 1981 ordinance was economically worthwhile loses its importance. It becomes more appropriate simply to thank those who enacted, enforced, and complied with it and to hope that we will take similar steps to safeguard the next generation.

Insurance

Losses suffered during natural disasters are borne primarily by three groups: the individuals who experience the disasters, private insurers, and governments. Government involvement can be in the form of state-issued insurance or of disaster relief. The role of insurance, especially in earthquake mitigation, is currently the subject of much debate.

The basic principle of insurance is simple: a large number of people exposed to some kind of risk pay premiums to a third party, such as a private insurance company. The company then pays the individuals who actually suffer a loss an amount sufficient to cover the loss, or some agreed fraction of it. The company sets the premiums both to cover the expected losses and to make a profit.

To set the rates accordingly, insurers attempt to determine the level of risk to which individual policyholders are exposed. Thus, some people pay much more for the same benefit than others. Older people, for example, or people with certain medical conditions, typically pay more for life insurance that do younger or healthier people. If insurers did not differentiate in this way, the system would work poorly, because people who knew that they were at high risk would buy insurance and those at low risk would not buy it, so premiums would escalate and the pool of insured people would decline to impractical levels. This phenomenon is termed *adverse selection*. Thus, insurance operates by market-driven principles that are quite different from the more socialist principles behind, say, a state-run welfare system, which is funded by universal taxation. Nevertheless, governments typically put pressure on the insurance industry to make it operate more like a welfare system.

For insurance against natural disasters, there is a real problem in determining the level of risk. These events are so infrequent, but so large in magnitude, that past experience, even averaged over many years, may be a poor guide to the future. As we discussed in earlier chapters, the frequency of earthquakes may change over decades or centuries. There's a widespread belief, for example, that earthquake activity in California is currently increasing after many decades of relative quiescence. Furthermore, population growth and development raise the likely losses, even without any change in the frequency of the natural events themselves.

During the 1980s and early 1990s, as several major earthquakes struck California, many insurance companies tried to get out of the business of offering earthquake coverage. The state responded by passing a law that forced any insurance company offering homeowners' insurance in the state to offer earthquake insurance as well. For a while, this worked. After the 1994 Northridge earthquake, however, when private insurance companies paid out $8.5 billion to settle residential insurance claims, matters came to a head. In 1996, several companies, led by Allstate Insurance, threatened to cease renewing California homeowners' policies altogether.

To deal with the crisis, the California legislature created a new agency, the California Earthquake Authority, which is a state-run program devoted solely to providing earthquake insurance. A problem

immediately arose concerning the setting of premiums: To what extent
should they reflect the variations in earthquake hazards in different
regions of the state?

The initial plans, put forward by insurance experts on the basis of
the available scientific information, were to establish large differences in
premiums. For example, homeowners in Sacramento, far from the most
dangerous faults, were to be charged an annual premium of $1.10 for
$1000 of coverage, while homeowners in Oakland, close to the
Hayward fault on the east side of the San Francisco Bay, were to be
charged 10 times as much—$11—for the same coverage.

Sensing that such steep differentials would be politically unacceptable, Insurance Commissioner Chuck Quakenbush ordered that rates
throughout the Bay Area be leveled to a fixed $5.25 per $1000. But this
still left a steep differential between the Bay Area and the Los Angeles
area, where the proposed rates averaged only $3.30. In 1997, under fur-
ther public pressure, rates in the Bay Area were lowered to about $4.60.

Scientists have been closely involved in the debate over insurance
rates. How certain is it that damaging earthquakes are becoming more
frequent in California? How certain is it that the San Francisco Bay
Area faces greater risks than southern California? With billions of dollars at stake, scientists have been pressed for clear answers, but these are
not always forthcoming. There simply isn't a complete consensus on
these questions. For example, geophysicist David Jackson of UCLA is
skeptical of our ability to determine precisely the probability of earthquakes at different locations, because future earthquakes may occur on
faults that have not yet been discovered (see Chapter 6). He also thinks
that the evidence for a major increase in the frequency of damaging
earthquakes in California is overstated. Thus, Jackson has testified at
rate-setting tribunals on behalf of consumers groups that oppose rate
increases, and he is skeptical of attempts to tailor insurance premiums to
the layout of known faults.

While respecting that point of view, our own opinion is that we
should do everything we can to make earthquake premiums reflect the
risk to which individual homeowners are exposed, even if the scientific findings are not yet a matter of universal agreement. We believe this
for two reasons. First, this is the only way to make the insurance system
viable in the long term. If homeowners in Oakland or in the Bay Area
generally perceive that they are getting a good deal, while homeowners in Los Angeles perceive that they are paying too much, then adverse

selection will kick in, and the sales of policies will decline and will become concentrated in the higher risk Bay Area, endangering the fiscal viability of the California Earthquake Authority. No doubt this trend will be reinforced by private insurers who will offer cut-rate policies to southern Californians, further undermining the state agency. Thus, the California Earthquake Authority, so long is it is in the insurance business, needs to adhere to the market principles that underlie the industry.

Second, and more important, insurance premiums should reflect real risks, as best we know them, in order to promote change. We must be aware that there are consequences when we choose to live in earthquake-prone areas. We must realize that it matters whether or not our homes are able to withstand the intensities of shaking that are likely to occur. The availability and cost of earthquake insurance is one effective tool to encourage this awareness. If the state blunts this tool by leveling premiums between high- and low-risk areas—or between classes of dwellings that differ in their vulnerability to shaking, liquefaction, or ground failure—then a major opportunity to persuade our citizens of the importance of earthquake mitigation will have been lost.

Not every loss can be foreseen or prevented, of course, even with the best scientific knowledge, the wisest government, and the most well-informed populace. There will always be an important place for federal, state, and private relief programs for the victims of earthquakes, volcanic eruptions, and other natural disasters. But policies that aim to restore people's losses no matter how foolhardy their choice of home-site, and no matter how few precautions they have taken to withstand Nature's fury, are likely themselves to lead to disaster, for no agency or government will be able to shoulder the resulting burden without enormous diversion of money and effort.

The increase in losses from natural disasters in the United States over the past three decades had been staggering. During the 1960s, total economic losses from natural disasters averaged only about $3 billion per year, but by the mid-1990s they were running at 10 times that level—a rate of increase that far outpaced inflation. An increase in the rate of occurrence of damaging events, such as earthquakes and hurricanes, may be partly to blame, but increasing population density and continuing willingness to locate homes and businesses in hazardous areas are more crucial. It is essential, in our view, that we mitigate the hazards by encouraging responsible development. Significant financial incentives, combined with educational efforts, could go a long way toward solving the problem.

Earthquakes in Los Angeles are frequent and will likely become more frequent in the twenty-first century. This circumstance is favorable to mitigation efforts, whether through insurance, through building codes, or through civil preparedness. In communities where destructive events are less frequent, mitigation efforts are more difficult to motivate. Human life is time-limited, and therefore all cost-benefit analyses, as well as our own instincts, lower the value of future benefits by a factor reflecting how far in the future they are likely to occur. A major earthquake in Boston or a large-scale eruption of Mount Rainier will surely happen one day, but probably not within the few decades with which we as individuals are most concerned. How we respond to these kinds of risks will say a lot about our commitment to our children and our children's children.

A Wider View

Although this book has focused on the United States, we cannot close without stressing how much more of a problem earthquakes and volcanoes present to many less affluent countries. Here is a quick comparison. The 1989 Loma Prieta earthquake (magnitude 7.1) caused losses equal to a mere 0.2 percent of the gross national product of the United States in that year. The 1986 San Salvador earthquake caused losses equal to 31 percent of the gross national product of El Salvador in that year. And the 1972 Managua earthquake (magnitude 5.6) caused losses equal to *40 percent* of Nicaragua's gross national product in that year. The message is simple: for the United States even large earthquakes, thus far, have been economic pinpricks; for poor countries, they can spell economic ruin.

Can developing countries afford to mitigate earthquake hazards? According to Andrew Coburn and Robin Spence, of Britain's Cambridge University, they cannot afford not to. At the request of the Turkish government, Coburn and Spence undertook a study of the earthquake problem in Turkey. Turkey has suffered more than 100 lethal earthquakes during the twentieth century, and about 75,000 people have died in them. Many of the deaths have occurred in the villages and small towns of eastern Turkey, where buildings have traditionally been constructed of weak, unreinforced masonry.

Coburn and Spence found that the installation of a single ring-beam around each building, at a total cost of about $200 million, would save about $250 million in reconstruction costs over 25 years and would also save about 2000 lives and a much larger numbers of injuries. With a larger capital investment (enough to install three ring-beams at different heights around each building), the economic savings would be even greater, and about 6500 lives would be saved. The case for carrying out these strengthening measures seems compelling. Similar arguments can be made for the value of mitigation efforts in many developing countries that are exposed to high seismic risks.

Yet several factors tend to impede mitigation programs. Capital (including foreign aid) is often scarce. Governments often ignore the needs of rural areas in favor of urban development. Traditional building practices are often difficult to change. And other natural disasters, such as floods and droughts, may loom larger than earthquakes or volcanoes in the national consciousness. Coburn and Spence emphasize the importance of seismic hazard education and of involving the rural population itself in the design and construction of improved buildings.

There are many ways in which scientific research, although largely conducted in the developed countries, can help to mitigate the effects of earthquakes and volcanoes around the globe. For example, it may become possible to monitor the level of the entire surface of the Earth continuously, to an accuracy of an inch or two, through various space-borne ranging techniques such as the laser interferometry technique mentioned in Chapter 5. Such a capability could allow us to detect the ground deformation that precedes volcanic eruptions and hence to provide advance warnings of eruptions in areas of the world that are rarely visited by geologists. Eventually, as the earthquake cycle becomes better understood, it may be possible to say something about seismic hazards through these same techniques.

Whatever the energy and resources we put into research and mitigation efforts, however, we have to accept that we cannot guarantee our safety here on Earth. We cannot prevent all casualties or all economic losses, even from the moderate-sized earthquakes and volcanic eruptions with which we are familiar. And as for the truly cataclysmic events—a caldera-forming eruption at Yellowstone or a sector collapse in Hawaii—there is basically nothing we can do at this point to protect ourselves from them. Luckily, such an event will probably not happen for 100,000 years or more; thus, until our science and technology have become equal to coping with these megacatastrophes, a healthy fatalism seems the only reasonable attitude toward them.

A Longer View

A hundred thousand years—more than a thousand human lifetimes laid end to end! And yet, in this time, the San Andreas fault will have slipped by a mere 2 miles—only one-hundredth the distance it has slipped over its own geological lifetime. And, for all its age, the San Andreas fault is a mere youngster. New York's Ramapo fault is 30 times older: over its half-billion-year lifetime it has seen ocean basins open up and close again, mountain ranges rise up and wear down, and countless species of living things evolve and become extinct. And yet the Earth is 10 times older than the Ramapo.

Geology has more to teach us than the practical importance of wall anchors, foundation bolts, and insurance policies. Like astronomy, it gives us a view of ourselves from the outside, of humankind in its natural context—a perspective so distant that individuals and nations and generations lose their identity in a fleet of onward-sailing humanity. The Sumerians are a little ahead, our remote descendants are a little behind, but we call out the hours of the watch to one another, and the hazards of the voyage, as we cross time's ocean. Let us take care that no storm disperses us.

Where our forebears wondered at the supernatural, we see Nature's own wonders and are no less inspired. Which is more uplifting: to spy a bloodthirsty demon on Mazama's peak or to trace the whole tangled web that leads from the Earth's beginnings to that great eruption—and to humans capable of marveling at it? What gives us the most purpose and hope in life: to dwell entirely in the here and now, to invent hereafters, or to know that each of us, though mortal, takes part in a timeless voyage of discovery?

APPENDIX: SOME
METHODS EXPLAINED

Earthquake Magnitudes

A number of magnitude scales have been designed over the years, and they each have their uses and shortcomings. In this book, we use a scale of magnitude called the *moment-magnitude scale*. This scale was developed by Caltech's Hiroo Kanamori in the mid-1970s to allow for the measurement of moderate to large earthquakes—those capable of causing serious damage. The better-known *Richter scale* is suitable for the measurement of small-to-moderate earthquakes, as measured by seismographs in the vicinity of the rupture. The term *Richter scale* has become so ingrained, however, that reporters and others often use the expression regardless of what scale has actually been used.

The moment-magnitude scale is a measure of the total energy radiated from the vicinity of the fault during the rupture. Energy is released because a large amount of rock jumps from a position in which it is under great stress to a new position in which it is under less stress; that is, from a higher to a lower energy state. Therefore, there is a surplus of energy, most of which is carried away by the shaking of the ground; that is, by the earthquake. As an analogy, consider dropping a pebble into a pond. The pebble moves from a (gravitationally) higher to a lower energy state, and the energy difference is carried away and dissipated largely by the resulting ripples on the water.

In the pebble example, the total amount of energy released can be calculated either from the pebble's fall (How big was the pebble? How far did it fall?) or from the resulting ripples (How high were they?). Similarly, we can estimate the energy released by an earthquake either from the geometry of the slippage or from the characteristics of the seismic waves recorded by seismographs at some distance from the fault.

To estimate the energy from the fault rupture, we need to know the area of the fault surface that has slipped and the average distance of the slip. The area of the rupture is the length of the ruptured segment multiplied by the depth of the rupture, which can be estimated from the distribution of aftershocks. (Most aftershocks occur somewhere near the portion of the fault plane that broke in the mainshock, so the depth of the deepest aftershocks gives an approximation of the depth of the main rupture.) The area of slippage is then multiplied by the average amount of slip (which, in the case of faults on land, can be measured from the displacement of surface features such as fences and roads) and by a factor called the *shear modulus of elasticity,* which tells us how much strain will result from the application of a given stress imposed on the rock. The resulting measure of total energy released by the rupture (called the *seismic moment*) is then converted to a logarithmic scale. A couple of "fudge factors" are thrown in to make the scale dovetail as neatly as possible with the Richter scale. The end result is the *moment magnitude,* which we refer to simply as *magnitude* (abbreviated as M). Each unit on the magnitude scale corresponds to an approximately thirtyfold increase in the total energy released in the earthquake.

To estimate magnitude from seismographic readings, seismologists measure the amplitude (size) of the recorded waves and then correct for the lessening of the amplitude that occurs with increasing distance from the location of the rupture. The measurements are made specifically on long-period (or low-frequency) waves, for two reasons. First, because most of the energy is carried by the long-period waves, only they can give accurate information about the total energy radiated from the fault. Second, long-period waves do not die out rapidly with distance, whereas short-period waves do. Therefore, only long-period waves can be used in the analysis of distant sources.

The fact that moment magnitude can be estimated in these two different fashions—from both the source and the effects—means that it can be used both for prehistoric earthquakes (if the details of the slip are preserved in the geological record) and for modern earthquakes that take place under instrumental surveillance.

Because the magnitude scale is a logarithmic scale, it is open at both ends: there is no such thing as an earthquake of the greatest or the least possible magnitude. On the Earth, of course, the magnitude of an earthquake is theoretically limited by the size of the Earth. If a fault going all the way around the planet were to rupture over its entire length and through the entire thickness of the brittle lithosphere, the resulting earthquake would have a magnitude of about 10.6, but bigger earthquakes could happen on bigger planets. The biggest earthquake ever recorded on Earth, the 1960 Chilean earthquake, produced by rupture of a 600-mile long segment of a subduction zone, had a magnitude of 9.5. The 1700 Cascadia subduction earthquake, as estimated from the Japanese tsunami records, had a magnitude of about 9.0. The great San Francisco earthquake of 1906 had a magnitude of about 7.7. Aftershocks of earthquakes, and the earthquakes caused by spontaneous slippages of small amounts of rock that go on all the time, can have magnitudes down to zero or into the negative numbers.

It should be emphasized that increasing magnitude does not translate in any simple way into more severe or longer lasting shaking at any particular location: what it does mean is that the severe shaking will be felt over a wider area. For example, if you were standing next to the San Andreas fault when it ruptured, it would make little difference to you whether the fault ruptured to a distance of 150 miles away from you or to twice that distance, because the high-frequency waves arriving from the farther reaches of the rupture would be too weak to add noticeably to your experience. On the other hand, structures such as large buildings or bridges, which are capable of responding to low-frequency seismic waves, *would* take more of a beating in the case of the longer rupture, because the low-frequency waves travel so well. Many other factors, such as the properties of the intervening rock and the local soil characteristics, greatly influence the intensity of shaking produced by an earthquake of a given magnitude, as discussed in Chapter 6.

Global Positioning System

The Global Positioning System (GPS) is a set of 21 military satellites orbiting the Earth at an altitude of about 12,000 miles. All the satellites continuously beam pseudorandom (noiselike) radio signals toward the Earth. A receiving device is placed at the spot whose precise location needs to be determined. The device records the signals from three or four satellites that happen to be somewhere in the sky overhead at the time. For each of the signals, the device tries to match the received signal to versions of the signal that have been shifted by various intervals of time (this is called auto-correlation analysis). The version that best fits the incoming signal tells the device the satellite-to-ground transit time of the signal, which can then be directly converted into distance. Knowing its precise distance to a single satellite (whose position is known), the device can locate itself on the surface of a sphere around the satellite. Knowing its distance to four satellites, it can locate itself to the unique point where the four spheres intersect. (In practice, it is sufficient to use signals from three satellites; this leads to two possible locations, but only one of them is on the Earth's surface.)

To prevent potential enemies from using the GPS to determine position with the same precision as the U.S. military, the Defense Department introduced a secret wobble into the signal. Scientists at the Massachusetts Institute of Technology, however, proved themselves to be excellent counterintelligence agents: they discovered that the wobble could be measured and subtracted by examining the signals received at a ground station whose position was already precisely known. In addition, it has proved possible to use the phase (wave-front position) of the signal's 19-centimeter (7-inch) carrier wave to refine the position estimate even further. Thus, while the Defense Department originally intended that civilians should be able to achieve an accuracy of no better than about 100 yards, geophysicists have attained accuracies of less than half an inch—good enough to detect rapid tectonic movements within a year or less.

Radiocarbon Dating

The radiocarbon dating technique, for which UCLA professor Willard Libby received the 1960 Nobel Prize in Chemistry, depends on measuring the relative abundance of two isotopes of carbon in the sample being studied: carbon-12, the common form, and carbon-14, a radioactive form that is present only in minute traces. Carbon-14 is generated high in the Earth's atmosphere when cosmic rays strike atoms of nitrogen-14, the common form of nitrogen that makes up most of the air we breathe. This carbon–14 is incorporated into carbon dioxide, which is then taken up by plants during photosynthesis (and thus indirectly by animals, too).

Carbon–14 decays back to nitrogen with a half-life of 5730 years. Thus, after a plant or animal dies, the ratio of carbon–14 to carbon–12 in its remains goes steadily down. The amount of carbon–12 in the sample is measured by ordinary chemical analysis. The amount of carbon–14 is measured in one of two ways. In the first method, the individual disintegrations of the carbon–14 atoms in the sample are counted in a special detector. To give accurate results, this method requires large samples and lengthy counting. The second method uses mass spectrometry: a small portion of the sample is ionized (the atoms are given an electric charge), and the atoms are "weighed" by measuring their acceleration in a magnetic field. Although either method can give reliable results, the radiocarbon technique as a whole is liable to several sources of error. In particular, the concentration of carbon–14 in the atmosphere has fluctuated over the centuries. For this reason, a given carbon–12/carbon–14 ratio may sometimes be consistent with dates as much as 200 years apart.

GLOSSARY

Italicized words within the definitions have their own glossary entries.

a'a (pron. AH-ah)   A *lava* flow whose surface is highly irregular and covered with rough blocks (contrast *pahoehoe*).

aftershock   An earthquake that occurs shortly after a larger earthquake and in the same region.

alkalic basalt   A kind of *basalt* that is enriched in the alkali metals, sodium and potassium.

andesite   A volcanic rock containing 54–62 percent *silica*.

anticline   Folded layers of rock or sediment with the fold flanks sloping away from each other. When youthful, may be associated with a rise in the terrain.

aseismic   Not producing detectable earthquakes.

ash   In volcanology, finely divided material produced during an explosive eruption. Unlike ash in the common sense, it is not the product of combustion.

asthenosphere   The portion of the Earth's *mantle*, about 320 miles thick, directly below the *lithosphere*. Although solid, the asthenosphere is more ductile than the rock of the lithosphere.

B.C.E. (Before the Common Era)   Politically correct term for B.C. (Before Christ).

bar   A measure of *stress*, approximately equivalent to atmospheric pressure at sea level.

basalt   A dark, fine-grained volcanic rock—the typical eruptive product of oceanic *shield volcanoes*. It contains 45–54 percent *silica*.

blind thrust fault   A *thrust fault* that does not reach the ground surface.

C.E. (Common Era)   Politically correct term for A.D. (Anno Domini).

caldera   A large bowl-shaped depression produced by the collapse of a volcano's summit in the course of a major eruption.

carbon-14 A *radioactive* isotope of carbon, which decays to nitrogen-14 with a *half-life* of 5730 years.

characteristic earthquake A repeating earthquake that involves a similar distribution of *slip* over the same fault segment on each occasion.

clamping stress (or normal stress) A *stress* exerted across the plane of a *fault*. A positive clamping stress tends to prevent slip on the fault.

creep *Slip* too slow to generate detectable seismic waves.

cross-correlation analysis A mathematical technique to establish the time difference between two similar waveforms.

crust The outer shell of the Earth, about 4 miles thick under the ocean and up to 30 miles thick under the continents. The crust and the outer portion of the *mantle* together make up the *lithosphere*.

dacite A volcanic rock containing 62–70 percent *silica*.

dendrochronology The dating of past events by the study of tree rings.

detachment A nearly horizontal *fault* separating an upper from a lower layer of the *crust*.

dike A sheetlike crack in rock opened up and filled by advancing *magma*.

dip The vertical component of a *fault* plane, or its slope.

dip-slip fault A *fault* that slips along its *dip* or slope.

Eocene epoch The geologic period from about 58 to 37 million years ago.

fault A fracture separating two bodies of rock, along which the rock masses have slipped against each other.

flood-basalt province A region covered by extraordinarily extensive and voluminous *basaltic* eruptions over a period of 1–2 million years. Believed to be formed when the head of a *mantle plume* reaches the *lithosphere*.

footwall Refers to the block below a sloping *fault*.

foreshock An earthquake that occurs shortly before a larger earthquake and in the same region.

fumarole A steam or gas vent, usually found in volcanic areas.

gabbro Dark crystalline rock produced by the solidification of *basaltic magma* at depth.

geyser A hot spring that erupts in intermittent jets.

global positioning system (GPS) The hardware and software used to determine one's position on Earth by measuring the time of arrival of radio signals transmitted from a constellation of U.S. Defense Dept. satellites.

granite In common usage, any crystalline, *quartz*-bearing rock that has solidified at depth.

hanging wall Refers to the block above a sloping *fault*.

harmonic tremor A sustained, low-frequency seismic oscillation associated with the rise of *magma* near a volcano.

Holocene The period since the end of the last ice age, roughly 10,000 years ago.

hot spot   The site of interaction between a *mantle plume* and the overlying *lithosphere,* characterized by intense volcanism.

howdah   A pavilion mounted on the back of an elephant.

hydrostatic pressure   The pressure exerted in all directions by a given point in a body of water, usually caused by the weight of water overlying it.

hydrothermal   Of a geological hot–water system.

inclination   The angle at which the Earth's magnetic field lines meet the Earth's surface. Inclination varies with latitude.

interferometry   The measurement of the relative position of objects by observing the interaction of coherent light or other waves emanating from them. Similar to *cross-correlation analysis,* but performed physically rather than mathematically.

isotope   Any one of a number of variants of an element characterized by different mass numbers but the same atomic number. Some isotopes are stable; others are unstable or *radioactive.*

lava   Molten volcanic rock above ground.

lava tube   A tube or tunnel within solid *lava,* along which still-molten lava is flowing or has flowed. After the eruption has ended, the tube may remain as an elongate cave.

left–lateral   Of a *strike-slip fault* in which each side of the fault, seen from the other side, slips to the left.

lithosphere   The relatively rigid outer shell of the Earth, about 60 miles thick, overlying the *asthenosphere.* The lithosphere includes the *crust* and the outer *mantle.*

load stress   The *stress* imposed by the weight of overlying rock.

logarithmic scale   A scale in which each unit increment corresponds to a constant proportionate (commonly tenfold) increase in the quantity measured.

magma   Molten rock underground.

magma chamber   A large *magma* reservoir, whose contents may erupt volcanically or may solidify in place, producing a *pluton.*

magnitude   A measure of the size of an earthquake. In this book, synonymous with *moment-magnitude.*

mainshock   The largest earthquake in a sequence of earthquakes.

mantle   The middle shell of the Earth's interior, between *crust* and *core.* It includes the *non-crustal lithosphere* (ca. 80 miles thick), the *asthenosphere* (ca. 320 miles thick), and the lower mantle (ca. 1300 miles thick).

metamorphic   Showing characteristic alterations in crystalline structure caused by long exposure to high temperature and/or pressure.

mid–ocean ridge   A range of undersea mountains, along whose central axis new oceanic *crust* is being formed.

Miocene epoch   The geologic period between about 24 and 5 million years ago.

moment magnitude   A *logarithmic* measure of the size of an earthquake, calculated from the dimensions of the *fault* rupture, and proportional to the total amount of energy released during the earthquake.

monocline   A single-sloped bend in strata that are otherwise flat-lying.

normal fault   A *dip-slip fault* in which the upper face (*hanging-wall* block) slips downward, relative to the lower face (*footwall* block).

overburden pressure   The pressure exerted by the weight of overlying rock.

pahoehoe (pron. pah-HOY-hoy)   A *lava* flow whose surface is smooth or ropy (contrast *a'a*).

paleomagnetism   The weak magnetism of volcanic rock, imposed by the Earth's magnetic field at the time the rock solidifies.

paleoseismology   The study of past earthquakes from the traces left by them in geologic landforms and layers.

pali   A Hawaiian cliff formed by *normal faulting* or by landsliding.

peat   A loose deposit of plant materials in process of carbonization.

plate, tectonic   A piece of the *lithosphere* that, to a first approximation, moves as a rigid unit across the Earth's surface.

plate tectonics   A conception of the Earth's *lithosphere* as a finite set of nearly rigid, independently moving plates.

Pleistocene epoch   The period between about 2 million and 10,000 years ago, notable for recurring ice ages.

Plinian column   A buoyant column of *pumice, ash,* and hot gas that is ejected explosively from a volcanic vent and then rises tens of thousands of feet into the atmosphere.

plumelet   A hypothesized subportion of a *mantle plume,* formed as it rises through the asthenosphere.

pluton   A large mass of rock formed by the solidification of *magma* underground.

potassium–argon dating   A technique to estimate the age of a rock by measuring the fraction of the potassium-40 in the sample that has decayed to argon-40.

primary (P) waves   The fastest moving seismic waves, consisting of forward-and-back compression and rarefaction of the rock.

pumice   A very light rock with a spongelike texture produced by the expansion of bubbles of gas within a magma and its explosive disarticulation.

pyroclastic flow (also ash flow, nuée ardente, or "glowing avalanche")   A mass of hot pumice and ash, buoyed by hot gas, that traverses the ground at high speed.

quartz   The most common form of crystalline *silica*.

radioactive   Prone to partial disintegration of nuclear structure, with resulting emission of radiation.

radiocarbon dating A technique to estimate the age of animal or plant remains by measuring the ratio of *carbon-14* to carbon-12 in the sample.

refraction A change in the direction of propagation of waves as they cross the boundary between two media that transmit waves at different speeds.

rejuvenation–phase volcanism Small eruptions that take place long after a volcano's *shield*-building period is over.

resurgent dome A dome formed by minor eruptions of viscous *lava* within a crater or *caldera,* in the aftermath of a major eruption.

reverse fault A *dip-slip fault* in which the upper face (*hanging-wall* block) moves upward relative to the lower face (*footwall* block).

rhyodacite A volcanic rock with a *silica* content intermediate between *dacite* and *rhyolite.*

rhyolite A light-colored volcanic rock containing 70–77 percent *silica.*

rift zone A zone of rifting that runs down the side of a volcano, along which flank eruptions occur.

rift A split in the Earth's crust, caused by extension.

right-lateral Of a *strike-slip fault* in which the each side of the fault, seen from the other side, slips to the right.

sand-blow (also sand-boil or sand volcano) A site where intense shaking has pressurized and liquefied the sandy subsoil and expelled it though a vent onto the surface.

seamount A mountain entirely beneath the ocean surface.

secondary (S) waves Seismic waves, slower moving than primary waves, consisting of side-to-side shearing motions of the rock.

sector collapse The collapse of a large sector of a volcano.

segment A portion of a *fault* bounded by some geometric or geological landmarks, or suspected of breaking as a unit.

seismography The recording and analysis of seismic waves.

seismology The study of earthquakes and the use of earthquake waves to study the structure of the Earth.

shear stress A *stress* tending to cause two adjacent parts of a solid to slide past each other.

shield volcano A volcano with gently sloping sides, formed from *basaltic lava* flows. The typical product of oceanic volcanism (contrast *stratovolcano*).

silica Silicon dioxide.

sinter Deposits that have formed in a *hydrothermal* system.

slip Movement between the two surfaces of a *fault.*

spreading center An alternative name for a *mid-ocean ridge,* emphasizing its role as the site of creation of new lithosphere. These zones are not always located in the middle of oceans, and they are even occasionally found on land. They are linear features, however, not points, as the word *center* suggests.

strain   The fractional change in the dimensions of a body induced by *stresses*.

stratigraphy   The study of layered sedimentary deposits.

stratovolcano   A steep (ideally conical) volcano composed of alternating layers of *lava-flow* and *pyroclastic-flow* deposits. The typical continental volcano, produced by the eruption of mostly medium-to-high-*silica magmas* (contrast *shield* volcano.)

stratum (pl. strata)   A layer of rock or soil deposited by sedimentation.

stress shadow   A temporary relief of *stress* over a wide area, caused by the rupture of a large *fault*.

stress   The force exerted on a unit area of a body.

strike   The direction or bearing of a horizontal line in the plane of a fault. It is perpendicular to the direction of the fault's *dip*.

strike–slip fault   A *fault* whose two surfaces slip horizontally, sideways, past each other during a rupture.

subduction   The recycling of old, usually oceanic, *lithosphere* back into the asthenosphere.

syncline   Folded layers of rock or sediment with the fold flanks dipping toward each other. When youthful, may be associated with subsidence of the terrain.

talus   Rocky debris at the base of an eroding cliff or steep slope.

tectonic   Having to do with the large-scale structure or architecture of the Earth's surface.

tephra   Fragmented rock that was thrown into the air during a volcanic eruption.

tholeiite (pron. TOE-lee-ite)   A common form of *basalt*, containing about 49 percent *silica*.

thrust fault   A *reverse fault* whose *dip* is closer to horizontal than to vertical.

transform fault   A *fault* connecting two segments of a *trench* or a *mid-ocean ridge* or connecting a trench with a ridge. At a transform fault, two adjacent *tectonic plates* slide horizontally past each other.

travertine   A form of calcium carbonate that has precipitated from a *hydrothermal system*.

trench   A trough in the ocean floor where old oceanic *lithosphere* begins its descent by *subduction* into the *mantle*.

tsunami   A fast-moving oceanic wave, or set of waves, caused by an earthquake or landslide (commonly but incorrectly referred to as a tidal wave).

tuff   Rock formed by consolidation of *pumice* or *ash*.

xenolith   A foreign or exotic rock fragment brought up with *magma*.

FOR MORE INFORMATION

General

Bolt, B. A. (1993). *Earthquakes.* New York: W. H. Freeman.

Decker, R., and B. Decker (1989). *Volcanoes.* New York: W. H. Freeman.

Fisher, R. V., G. Heiken, and J. B. Hulen (1997). *Volcanoes: Crucibles of Change.* Princeton, N.J.: Princeton University Press

Moores, E. M., ed. (1990). *Shaping the Earth: Tectonics of Continents and Oceans* (Readings from *Scientific American Magazine*). New York: W. H. Freeman.

Yeats, R., K. Sieh, and C. Allen (1997). *The Geology of Earthquakes.* New York: Oxford University Press.

http://quake.wr.usgs.gov/ (USGS seismology site).

http://www.yahoo.com/Science/Earth_Sciences/Geology_and_Geophysics/Volcanology/ (Yahoo's volcano section).

http://www.yahoo.com/Science/Earth_Sciences/Geology_and_Geophysics/Seismology/ (Yahoo's earthquake section).

Chapter 1: When Push Comes to Shove: Giant Earthquakes in the Pacific Northwest

Allègre, C. (1988). *The Behavior of the Earth: Continental and Seafloor Mobility.* Cambridge, Mass.: Harvard University Press.

Atwater, B., et al. (1995).Summary of coastal geologic evidence for past great earthquakes at the Cascadia subduction zone. *Earthquake Spectra* 11:1–18.

Satake, K., K. Shimazaki, T. Yoshinobu, and K. Ueda (1996). Time and size of a giant earthquake in Cascadia inferred from Japanese tsunami records of January 1700. *Nature* 379:246–249.

Chapter 2: Blasts from the Past:
Mount St. Helens and Her Sleeping Sisters

Bacon, C. R. (1983). Eruptive history of Mount Mazama and Crater Lake caldera, Cascade Range, USA. *Journal of Volcanology and Geothermal Research* 18:57–115.

Clark, E. E. (1958). *Indian Legends of the Pacific Northwest.* Berkeley: University of California Press.

Harris, S. L. (1988). *Fire Mountains of the West: The Cascade and Mono Lake Volcanoes.* Missoula, Montana: Mountain Press.

Lipman, P. W., and D. R. Mullineaux, eds. (1981). *The 1980 Eruptions of Mount St. Helens, Washington.* USGS Professional Paper 1250. Washington, D.C.: U.S. Government Printing Office.

Rosenfeld, C., and R. Cooke (1982). *Earthfire: The Eruption of Mount St. Helens.* Cambridge, Mass.: MIT Press.

Saarinen, T. F., and J. L. Sell *(1985). Warning and Response to the Mount St. Helens Eruption.* Albany: State University of New York Press.

U.S. Geodynamics Committee, National Research Council (1994). *Mount Rainier: Active Cascade Volcano.* Washington, D.C.: National Academy Press.

http://vulcan.wr.usgs.gov/ (Cascades Volcano Observatory home page).

Chapter 3: The Great Divide: Discovering the San Andreas Fault

Hill, M. L. (1981). San Andreas fault: History of concepts. *Geological Society of America Bulletin* 92:112–131.

Lawson, A. C., and H. F. Reid, eds. (1908, 1910). *The California Earthquake of April 18, 1906: Report of the State Earthquake Investigation Commission.* 2 vols. Washington, D.C.: The Carnegie Institution.

Powell, R. E., R. J. Weldon, II, and J. C. Matti, eds. (1993). *The San Andreas Fault System: Displacement, Palinspastic Reconstruction, and Geologic Evolution.* Geological Society of America Memoir 178. Boulder, Colorado: Geological Society of America.

Wallace, R. E., ed. (1990). *The San Andreas Fault System, California.* USGS Professional Paper 1515. Washington, D.C.: U.S. Government Printing Office.

Chapter 4: In the Trenches: Unearthing the Seismic History of the San Andreas Fault

Grant, L. B., and K. Sieh (1994). Paleoseismic evidence of clustered earthquakes on the San Andreas fault in the Carrizo Plain, California. *Journal of Geophysical Research* 99:6819–6841.

Jacoby, G. C., Jr., P. R. Sheppard, and K.E. Sieh (1988). Irregular occurrence of large earthquakes along the San Andreas fault: Evidence from trees. *Science* 241:196–199.

Sieh, K., M. Stuiver, and D. Brillinger (1989). A more precise chronology of earthquakes produced by the San Andreas fault in southern California. *Journal of Geophysical Research* 94:603–623.

Chapter 5: Lying in Wait: The San Andreas Fault Today and Tomorrow

Ellsworth, W. L. (1995). "Characteristic Earthquakes and Long-Term Earthquake Forecasts: Implications of Central California Seismicity." In *Urban Disaster Mitigation: The Role of Science and Technology.* F.Y. Cheng and M. S. Sheu, eds. 1–14. Amsterdam: Elsevier.

Harris, R. A., and R. W. Simpson (1996). In the shadow of 1857: The effect of the great Ft. Tejon earthquake on subsequent earthquakes in southern California. *Geophysical Research Letters* 23:229–232.

Sieh, K. (1996). The repetition of large-earthquake ruptures. *Proceedings of the National Academy of Sciences of the USA* 93:3764-3771.

Ward, S. N. (1996). A synthetic seismicity model for southern California: Cycles, probabilities, and hazard. *Journal of Geophysical Research* 101:22,393-22,418.

Working Group on California Earthquake Probabilities (1995). Seismic hazards in southern California: Probable earthquakes 1994–2024. *Bulletin of the Seismological Society of America* 85:22379–22439.

Chapter 6: The Enemy within the Gates: Earthquakes on Urban Faults

Bucknam, R. C., E. Hemphill-Haley, and E. B. Leopold (1992). Abrupt uplift within the past 1700 years at southern Puget Sound, Washington. *Science* 258:1611–1614 (See also the four following papers in the same issue.)

Dolan, J. F., et al. (1995). Prospects for larger or more frequent earthquakes in the Los Angeles metropolitan region. *Science* 267:199–205.

Heaton, T. H., et al. (1995). Response of high-rise and base-isolated buildings to a hypothetical Mw 7.0 blind thrust earthquake. *Science* 267:206–211.

USGS (1996). *USGS Response to an Urban Earthquake: Northridge '94.* USGS Open File Report 96-263. Washington, D.C.: U.S. Government Printing Office.

USGS and Southern California Earthquake Center (1994). The magnitude 6.7 Northridge, California, earthquake of 17 January 1994. *Science* 266:389–397.

Wald, D. J., T. H. Heaton, and K.W . Hudnut (1996). The slip history of the 1994 Northridge, California, earthquake determined from strong-motion, teleseismic, GPS, and leveling data. *Bulletin of the Seismological Society of America* 86:S49–S70.

http://www.usc.edu/dept/Earth/quake (Southern California Earthquake Center site).

http://www.dot.cagov/dist4/sfobbrto.htm (CalTrans San Francisco–Oakland Bay Bridge Seismic Retrofit site).

Chapter 7: The Little Volcano That Couldn't: Fear and Trembling at Mammoth Lakes

Hill, D. P. (1996). Earthquakes and carbon dioxide beneath Mammoth Mountain, California. *Seismological Research Letters* 67:8–15.

Hill, D. P., R. A. Bailey, and A. S. Ryall (1985). Active tectonic and magmatic processes beneath Long Valley caldera, eastern California: An overview. *Journal of Geophysical Research* 90:11,111-11,120.

Sieh, K.E., and M. I. Bursik (1986). Most recent eruption of the Mono Craters, eastern central California. *Journal of Geophysical Research* 91:12,539-12,571.

Chapter 8: Inland Murmurs: Earthquakes in the Basin and Range Province

Hodgkinson, K. M., R. S. Stein, and G. C. P. King (1996). The 1954 Rainbow Mountain–Fairview Peak–Dixie Valley earthquakes: A triggered normal faulting sequence. *Journal of Geophysical Research* 101:25,459-25,471

Machette, M. M., et al. (1991). The Wasatch fault zone, Utah: Segmentation and history of Holocene earthquakes. *Journal of Structural Geology* 13:137–149.

McPhee, J. (1980). *Basin and Range.* New York: Farrar, Straus, Giroux.

USGS (1964). *The Hebgen Lake, Montana, Earthquake of August 17, 1959: USGS Professional Paper 435.* Washington, D.C.: U.S. Government Printing Office.

Wernicke, B. (1992). "Cenozoic Extensional Tectonics of the U.S. Cordillera." In *The Geology of North America*. Vol. G-3. 553–581. Boulder, Colorado: Geological Society of America.

Yochelson, E. L., ed. (1980). *The Scientific Ideas of G. K. Gilbert*. Special Paper 183. Boulder, Colorado: Geological Society of America.

Chapter 9: On the Hot Spot: The Volcanoes at Yellowstone

Duncan, R.A., and M. A. Richards (1991). Hot spots, mantle plumes, flood basalts, and true polar wander. *Review of Geophysics* 29:31–50.

Fournier, R. O., et al. (1994). *A Field-Trip Guide to Yellowstone National Park, Wyoming, Montana, and Idaho—Volcanic, Hydrothermal, and Glacial Activity in the Region*. U.S. Geological Survey Bulletin 2099. Washington, D.C.: U.S. Government Printing Office.

Johnston, S. T., et al. (1996). Yellowstone in Yukon: The Late Cretaceous Carmacks Group. *Geology* 24:997–1000.

Pierce, K. L., and L. A. Morgan (1992). "The Track of the Yellowstone Hot Spot: Volcanism, Faulting, and Uplift." In *Regional Geology of Eastern Idaho and Western Wyoming*. Geological Society of America Memoir 179. P. K. Link, M. A. Kuntz, and L. B. Platt, eds. Boulder, Colorado: Geological Society of America.

Smith, R. B., and L. W. Braile (1994). The Yellowstone hot spot. *Journal of Volcanology and Geothermal Research* 61:121–187.

Chapter 10: A River Runs through It: The Mississippi and the New Madrid Earthquakes

Crone, A. J., and K.V. Luza (1990). Style and timing of Holocene surface faulting on the Meers fault, southwestern Oklahoma. *Geological Society of America Bulletin* 102:1–17.

Johnston, A. C., and E. S. Schweig (1996). The enigma of the New Madrid earthquakes of 1811–1812. *Annual Review of Earth and Planetary Sciences* 24:339–384.

Hildenbrand, T. G., and J. D. Hendricks (1995). *Geophysical Setting of the Reelfoot Rift and Relations between Rift Structures and The New Madrid Seismic Zone*. U.S. Geological Survey Professional Paper 1538-E. Washington, D.C.: U.S. Government Printing Office.

Kelson, K. I., et al. (1996). Multiple late Holocene earthquakes along the

Reelfoot fault, central New Madrid seismic zone. *Journal of Geophysical Research* 101:6151–6170.

Penick, J., Jr. (1976). *The New Madrid Earthquakes of 1811-1812*. Columbia, Missouri: University of Missouri Press.

Tuttle, M. P., et al. (1996). Use of archaeology to date liquefaction features and seismic events in the New Madrid seismic zone, central United States. *Geoarchaeology* 11:451–480.

Chapter 11: Along the Eastern Sea:
Earthquakes on the Atlantic Coast

Dutton, C. E. (1889). "The Charleston Earthquake of August 31, 1886." In *Ninth Annual Report of the United States Geological Survey to the Secretary of the Interior, 1887–'88*. J. W. Powell, ed. Washington, D.C.: U.S. Government Printing Office.

Ebel, J. E. (1996). The seventeenth century seismicity of northeastern North America. *Seismological Research Letters* 67:51–68.

Ebel, J. E., and A. L. Kafka (1991). "Earthquake Activity in the Northeastern United States. In *Neotectonics of North America*. D. B. Slemmons et al., eds. Boulder, Colorado: Geological Society of America.

Kafka, A. L., E. A. Schlesinger-Miller, and N. L. Barstow (1985). Earthquake activity in the greater New York City area: Magnitudes, seismicity, and geologic structures. *Bulletin of the Seismological Society of America* 75:1285–1300.

Marple, R. T., and P. Talwani (1993). Evidence of possible tectonic upwarping along the South Carolina coastal plain from an examination of river morphology and elevation data. *Geology* 21:651–654.

Obermeier, S. F. (1989). *The New Madrid Earthquake: An Engineering-Geologic Interpretation of Relict Liquefaction Features*. U.S. Geological Survey Professional Paper 1336-B. Washington, D.C.: U.S. Government Printing Office.

Chapter 12: Pele's Wrath: The Volcanoes of Hawaii

Decker, R. W., T. L. Wright, and P. H. Stauffer (1987). *Volcanism in Hawaii*. USGS Professional Paper 1350. 2 vols. Washington, D.C.: U.S. Government Printing Office.

Hawaii Scientific Drilling Project Team (1996). Hawaii Scientific Drilling Project: Summary of preliminary results. *GSA Today* (August):1–8.

Hazlett, R. W. (1987). *Geological Field Guide, Kilauea Volcano.* Volcano, Hawaii: Hawaiian Volcano Observatory.

Hazlett, R. W., and D. W. Hyndman (1996). *Roadside Geology of Hawaii.* Missoula, Montana: Mountain Press.

Heliker, C. (1993). *Volcanic and Seismic Hazards on the Island of Hawaii.* Honolulu: Bishop Museum Press.

Ihinger, P. D. (1995). Mantle flow beneath the Pacific Plate: Evidence from seamount segments in the Hawaiian-Emperor chain. *American Journal of Science* 295:1035–1057.

Moore, J. G., et al. (1989). Prodigious submarine landslides on the Hawaiian Ridge. *Journal of Geophysical Research* 94:17,465-17,484.

Rhodes, J. M., and J. P. Lockwood, eds. (1995). *Mauna Loa Revealed: Structure, Composition, History, and Hazards.* Washington, D.C.: American Geophysical Union.

Wright, T. L., T. J. Takahashi, and J .D. Griggs (1992). *Hawai'i Volcano Watch: A Pictorial History, 1779–1991.* Honolulu: University of Hawaii Press and Hawaii Natural History Association.

http://www.soest.hawaii.edu/GG/HCV/loihi.html (University of Hawaii's Loihi site).

http://www.soest.hawaii.edu/GG/HCV/kilauea.html (Hawaii Volcano Observatory's Kilauea site).

Chapter 13: The Years of Living Dangerously: How Do We Survive on an Earth in Turmoil?

Coburn, A., and R. Spence (1992). *Earthquake Protection.* New York: John Wiley and Sons.

Sarin, R. K. (1983). A social decision analysis of the earthquake safety problem: The case of existing Los Angeles buildings. *Risk Analysis* 3:35–50.

INDEX